Springer Aerospace Technology

The series explores the technology and the science related to the aircraft and spacecraft including concept, design, assembly, control and maintenance. The topics cover aircraft, missiles, space vehicles, aircraft engines and propulsion units. The volumes of the series present the fundamentals, the applications and the advances in all the fields related to aerospace engineering, including:

- structural analysis,
- aerodynamics,
- aeroelasticity,
- aeroacoustics,
- flight mechanics and dynamics,
- orbital maneuvers,
- avionics,
- systems design,
- materials technology,
- launch technology,
- payload and satellite technology,
- space industry, medicine and biology.

The series' scope includes monographs, professional books, advanced textbooks, as well as selected contributions from specialized conferences and workshops.

The volumes of the series are single-blind peer-reviewed.

To submit a proposal or request further information, please contact: Mr. Pierpaolo Riva at pierpaolo.riva@springer.com (Europe and Americas) Mr. Mengchu Huang at mengchu.huang@springer.com (China)

The series is indexed in Scopus and Compendex

Pavel Pavlovich Hachikyan

Countering Air Terrorism

Springer

Pavel Pavlovich Hachikyan
Candidate of Technical Sciences (Safety
in Emergency Situations), Researcher
HSE University
Moscow, Russia

ISSN 1869-1730 ISSN 1869-1749 (electronic)
Springer Aerospace Technology
ISBN 978-981-19-6171-7 ISBN 978-981-19-6172-4 (eBook)
https://doi.org/10.1007/978-981-19-6172-4

This Springer imprint is published by the registered company Springer Nature Singapore Pte Ltd.
The registered company address is: 152 Beach Road, #21-01/04 Gateway East, Singapore 189721, Singapore

Preface

The reason for the appearance of this book was the lack of currently relevant fundamental mechanisms for countering terrorism in aviation, as well as the growth of terrorist threats associated with the unstable world political situation. The use of aviation infrastructure facilities—airports and aircraft by terrorist organizations as targets of their illegal actions forces the world community to take more seriously the issue of ensuring anti-terrorist security in air transport, which determines the relevance of the book. At the same time, it is obvious that the possibility of committing terrorist acts is a consequence of serious shortcomings in the organization of anti-terrorist security in an aviation enterprise and, in particular, during inspection procedures that allow terrorists to freely carry prohibited items and substances both into the airport territory and directly on board the aircraft. The issue of anti-terrorist security becomes «No. 1» in the organization of the functioning of each aviation enterprise.

The main purpose of this book is to demonstrate the mathematical and mathematical-psychological models and methods developed by the author to reduce the risk of terrorist threats in an aviation enterprise by improving the organization of production processes and optimizing the distribution of financial resources when implementing systemic measures in relation to individual organizational measures of the anti-terrorist security system of an aviation enterprise, which determines the area of issues considered in it.

The book examines the issues of ensuring the anti-terrorist security of an aviation enterprise, provides an excursion into the history of terrorism in the aviation sector, and analyzes the legal regulation. The author provides a demonstration of the new approaches developed by him—mathematical and mathematical-psychological models and methods of reducing the risk of terrorist threats, which makes it possible to objectively increase the safety of air transport facilities, reduce the likelihood of terrorist attacks, save human lives and prevent significant material losses. The solutions of general humanistic problems proposed by the author of the book—preventing accidents and preserving the world—are an original feature of the book, which allows one to treat it with the necessary attention and interest for the reader. The concept of anti-terrorist security of an aviation enterprise, developed by the author of the book, has been widely tested and introduced into the activities of leading aviation enterprises, including the Interstate Aviation Committee (IAC), and is used in the investigation of aviation accidents.

The book is written in an understandable scientific language, contains a sufficient amount of explanatory material and formulas, will undoubtedly be useful both for a wide range of persons associated with ensuring aviation transport security and for the end users of airlines—passengers, in order to expand knowledge about the reliability of using air transport.

Moscow, Russia Pavel Pavlovich Hachikyan

Acknowledgements

The author of the book expresses gratitude to his scientific mentor, Honored Worker of Science and Technology of the Russian Federation, Professor, Doctor of Physical and Mathematical Sciences Kozlov Anatoly Ivanovich. Words of appreciation also dedicated to reviewers, respected experts in the field of security—Akinshin Ruslan Nikolaevich, Professor, Doctor of Technical Sciences and Anatoly Valeryevich Rybakov, Professor, Doctor of Technical Sciences.

Very special words of gratitude apply to my wife Joanna and daughter Emma, who always understood the fact, that I am often busy with scientific work.

Words of gratitude should be said to the team of Springer publishing house, which allows scientists around the world to disseminate and make available to everyone the most important scientific knowledge.

This work is an output of a research project «Study of models and methods of decision making under conditions of deep uncertainty» implemented as part of the Basic Research Program at the National Research University Higher School of Economics (HSE University).

With best wishes to the readers,
Pavel Pavlovich Hachikyan
Candidate of Technical Sciences (Safety in Emergency Situations), Researcher
Moscow, Russian Federation

Introduction

Relevance

The big resonance in the world is caused by tragic events on September 11, 2001 in the USA. Acts of terrorism, perfect members of the Al-Qaeda terrorist organization * by means of hijacking of aircrafts, claimed the lives not only of passengers but also the people in World Trade Center and the building of the Pentagon—all 2977 people died, 24 more persons remained in lists of missing persons. Events formed on September 11 a basis for revision of the security systems existing at that time and counteraction to terrorism of all objects of aviation infrastructure in the world, demonstrated their former low efficiency.

On August 24, 2004 at an interval of one minute, in air, the airliners Tu-134A-3 and Tu-154B-2 making flight from the Moscow Domodedovo Airport to Volgograd and Sochi blew up. The terrorist attacks conducted by suicide bombers claimed the lives of 89 passengers and crew members of airliners. The lack of necessary control of the organization of customs procedures of passengers and baggage from responsible persons of the airport gave an opportunity to terrorists to bring on board improvised explosive devices. Responsibility for the event is undertaken by the Karachay Jamia terrorist organization *.

On January 24, 2011, act of terrorism happened in the hall of the international arrivals of the Moscow Domodedovo Airport. The explosion made by the suicide bomber claimed the lives of 37 people, 172 wounded. Responsibility for actions is undertaken by the Islamic terrorist organization *Caucasus Emirate* *.

On October 31, 2015, there was the largest plane crash, which became the most mass death of citizens of Russia for all history of world aircraft—explosion of the passenger Airbus airliner of 321 Russian airlines *Kogalymavia* in air over the central part of the Sinai Peninsula. Drawdown of the improvised explosive device, which was mounted in a compartment of oversized baggage in a tail part of the liner served as a cause of the crash. Act of terrorism claimed the lives of 224 people. As a result of investigation, it was established that a terrorist attack is conducted by members of the Vilayat-Sinai terrorist group *, being ISIL branch * on the Sinai Peninsula.

Considering high costs of implementation of measures of ensuring acceptable level of safety, the scientific problem of decrease in risk of realization of terrorist threats in airline by improvement of the organization of productions and optimization of distribution of a financial resource at realization of system measures in relation to separate organizational measures of system of anti-terrorist safety of the air enterprise is staticized.

The solution of this scientific task in relation to separate organizational measures of system of anti-terrorist safety of airline is the purpose of the real research, as defined relevance.

The goal of a research achieved by a solution the author of the main scientific objectives

- Distributions of a limited financial resource between elements of highly reliable system of ensuring anti-terrorist safety by criterion of ensuring the maximum probability of protection against terrorist threats.
- Distributions of the set reliability between elements of highly reliable system of ensuring anti-terrorist safety by criterion of use of the minimum financial resource.
- Estimates of probability of realization of act of terrorism depending on quantity and reliability of boundaries of protection.
- Modeling of «system of ensuring anti-terrorist safety of airline».
- Developments of a method of selection of the experts involved in ensuring anti-terrorist safety of airline in qualification qualities.
- Developments of methodical recommendations about selection of the experts involved in ensuring anti-terrorist safety of airline in personal qualities.

In the Book, the Models and Methods Developed by the Author are in Detail Described

- Models of distribution of the set reliability between elements of highly reliable system of ensuring anti-terrorist safety by criterion of use of the minimum financial resource.
- Models of distribution of a limited financial resource between elements of highly reliable system of ensuring anti-terrorist safety by criterion of ensuring the maximum probability of protection.
- Model of assessment of probability of realization of act of terrorism depending on quantity and reliability of boundaries of protection.
- Model of «system of ensuring anti-terrorist safety of airline»
- A method of selection of the experts involved in ensuring anti-terrorist safety of airline in qualification qualities.
- Methodical recommendations about selection of the experts involved in ensuring anti-terrorist safety of airline in personal qualities.

The Scope of Scientific Research

The zone of responsibility of airline connected with ensuring protection against terrorist threats enters the borders of a research. A framework of a research is limited to development of the new scientific methods and models directed to decrease in risk of realization of terrorist threats in airline. As a basis of prevention of possible threat, the technical, organizational, and personnel-administrative measures (application of new methods, models) that beforehand carried out by airline directed to the prevention of terrorist threat are considered.

Theoretical and Practical Importance

The theoretical importance of a research consists of development of application of a mathematical apparatus at the solution of tasks of development of methods of staffing and modeling of distribution of a resource in relation to the solution of tasks in the scientific fields of ensuring aviation safety, the organization of production and operation of air transport.

The practical importance of a research consists of decrease in risk of realization of terrorist threats in airline that will objectively allow preventing possible consequences of terrorist attacks in the form of the human victims and material losses.

Methodology and Methods of a Research

The methodological basis of a research made: general scientific method of knowledge, legal and historical and legal methods, system and structural approach and system analysis, comparative, theoretical methods and idealization method. For the solution of the tasks of distribution of resources set in a research the mathematical apparatus and mathematical modeling (deterministic and statistical models), Neumann-Pearson's criterion and the Monte Carlo method used. Development of a qualification method of selection of shots is carried out with application of mathematical statistics and modeling. When developing methodical recommendations about a personal method of selection of shots the logic algebra, Boolean algebra, V. A. Lefebvre's, SMIL (MMPI) statistical scientific base theories are used. For check of adequacy of the developed methodical recommendations of selection of shots about personal qualities, used empirical and comparative methods of researches.

Use and Introduction of Results of Work

Results of research work of the author successfully used and introduced in activity:

- The Interstate Aviation Committee (IAC)—at investigation of the aviation incidents, for assessment of efficiency undertaken by airline of measures for ensuring anti-terrorist safety and compliance of level of training of personnel of service of aviation safety of airline to the tasks assigned to them;
- LLC GOSOBORONZAKAZ (Russian Federation)—when developing the concept of airline of civil aviation of the increased anti-terrorist security;
- The Russian academy of national economy and public service (RANEPA) under the President of the Russian Federation (Russian Federation)—when carrying out educational process in «Economic security», when studying by students of disciplines: national security, management of the organization (enterprise), control and audit, law enforcement agencies, psychology.

When carrying out a research author's inventions and useful models on devices to increase in reliability of a security system applied: the guard cabin, patent for the invention of RU 2586992 C1; the observation and watchtower, the patent for the useful RU 163226 U1 model.

Contents

Chapter 1
Terrorist Threats on Air Transport

1.1 Historical Digression to a Perspective

Infrastructure facilities of air transport and aircrafts are places of mass stay of people that give special, importance to questions of safety.

The concept "anti-terrorist safety" of aircraft is rather new, is under construction on the general principles of ensuring aviation safety, has the characteristics regulated by the current legislation. Harmonious and successful interaction of all links of a security system of the air enterprise is possible only at qualitative understanding of a role of each of its participants in which base the reliable legal foundation laid.

The concept "aviation safety" has several conventional definitions: according to the appendix 17 to the Convention on the international civil aviation of International Civil Aviation Organization, it—"a package of measures and also the human and material resources intended for protection of civil aviation against acts of illegal intervention" [37]. The air code of the Russian Federation provides a few other definition of aviation safety: "a condition of security of aircraft from illegal intervention in activity in the field of aircraft" [19]. Nevertheless, it is impossible to tell that these definitions fully open this concept, so, according to the author, most fully the concept "aviation safety" can be characterized as "the condition of security of the air enterprise from acts of illegal intervention in their activity, including terrorist orientation providing acceptable risk level of such intervention". Anti-terrorist safety in the concept of the book considered as an element of aviation safety, direct defining threats to security of airline from terrorists.

The concept "terrorism" (from Latin "terror", it is translated as fear, horror) can be characterized as the ideology of violence and aggressive impact on acceptance of the decision of public authorities, local governments or the international organizations connected with intimidation of the population or other forms of illegal violent acts [23]. The concept "terrorism", unlike rather modern—"aviation safety", "safety of flights", is known already for a long time: the first mentions of actions that can be described as terrorism appeared at the time of Ancient Rome at first as the description

P. P. Hachikyan, *Countering Air Terrorism*, Springer Aerospace Technology,
https://doi.org/10.1007/978-981-19-6172-4_1

of emotional exclamation ("Fear, horror!") for drama events [50], and further and various hostile bloody actions and incidents.

Fundamental problem of counteraction to terrorism from the state is the creation of reliable system of protection of citizens, society, and authorities from threat of terrorist intervention, at the same time, a basis of such counteraction is both precautionary measures, and the actual destruction of threats in case of their emergence and unwillingness to be given—in the modern world practically all states do not conduct negotiations with terrorists for the purpose of payment for release of hostages and performance of other conditions (of course, only if it formally is not connected with the need to receive extra time for the organization of storm and taking of terrorists). The ideology of the similar attitude of the states towards terrorists can be described in the words of the famous French writer and philosopher Zh. Zh. Russo: "To buy world from the enemy—means to supply it with means for new war" [48].

It is necessary to distinguish the concepts "terrorism" and "diversion" which, at first sight, can mistakenly seem identical. The concept "diversion" (from Latin "diversio", it is translated as a deviation, derivation) means the acts planned by criminals and directed to commission of explosion, the arson or other actions directed to destruction or damage of the enterprises, constructions, objects of transport infrastructure, means of communication, critical public infrastructure for undermining economic security and defense capability of the country [18]. These illegal acts are carried out is reserved and secretly, in difference from open fighting, however, propagandize the similar purpose—causing damage to the state opponent. Let's consider distinctions within the legislation of the Russian Federation: responsibility for diversion is provided by article 281 of the Criminal Code of the Russian Federation whereas responsibility for commission of terrorist actions is established by separate article 205 of the Criminal Code of the Russian Federation. In spite of the fact that both the specified criminal actions treat crimes against public safety and, in general, they are similar to extent of mixture, differentiation of diversion and terrorism needs to be carried out on the subjective party—the purpose of commission of crime. At sabotages, the purpose is—undermining economic security and defense capability of the country by removal from a system of key economic and defensive objects whereas in a case with terrorism—the purpose is violation of public safety, intimidation of the population and acceptance by authorities of the decision necessary to terrorists [18]. It should be noted that diversion on the specifics is well-planned secret military operation at which maintaining anonymity or deception concerning origin of threat—is the highest level of skill of carrying out diversion. In opposition to this statement, terrorism—does not happen anonymous. Particular persons or terrorist communities that undertake responsibility for deeds are responsible for acts of terrorism and also announce the purposes of their commission, show new threats and conditions under which similar acts of terrorism can repeat. Also, special services of the foreign states can be subjects of terrorist activity [47]. Similar terrorist attacks are conducted by them on behalf of a certain terrorist organization, at the same time, a main objective of such actions is in light, favorable to the state aggressor, to display certain political motives and to destabilize a situation of the affected state. It is also worth considering the possibility that in some countries, terrorist attacks can be conducted by individuals (or according to their instruction) who assume to benefit further from allocation of additional financial means by the budget of the state on fight against terrorism.

The relevance of the purpose of decrease in risk of realization of terrorist threats in airline is caused by special appeal of air transport to terrorists which consists in an opportunity how at the same time to claim many people lives, and to cause significant material damage, to cause the actions a serious public response and panic in society. For confirmation of relevance of the purpose is necessary to conduct the analysis and a research of the reasons of the most considerable aviation incidents connected with terrorism in recent years (Table 1.1).

The analysis of the reasons for the realized terrorist threats given in the table allows to draw a conclusion that the most often terrorist attacks are conducted because of essential shortcomings of the organization of safety, in particular: at unqualified holding customs procedures of passengers and baggage and also because of imperfection technical decisions in the field of ensuring examination and safety of an object of aviation infrastructure. Cases of exclusively realized acts of terrorism as similar examples allow to judge degree of importance of a perspective are considered by the author, to objectively represent the possible irreplaceable losses expressed by number of the victims. Similar statistics, but not information on the prevented terrorist attacks, allows to draw a conclusion on relevance of a research objective and need to continue to undertake and improve the measures directed to decrease in risk of realization of terrorist threats in airline.

It is necessary to pay attention that among the listed aviation incidents at which destructive impact on the aircraft was made from the outside—with use of surface-to-air missile systems is also specified. Ways of decrease in risks of similar incidents go beyond a framework of the real author's research, objectively are out of a zone of responsibility of officials of airline, on condition of implementation of flights in peace time, in the airspace allowed for civil flights.

Concretizing a range of terrorist threats in the field of civil aviation, it should be noted enough destructive terrorist ways and types of influence both on transport, and on social infrastructure among which: explosions and arsons with use of fire and cold weapon, weapons of mass destruction (radioactive and chemicals, biological weapon), taking of hostages, cyberterrorism, use of unmanned aerial vehicles as the instrument of terror and also various ways of psychological and information influence. Each of types of terrorist attack bears destructive consequences in the form of human victims and considerable material losses (Scheme 1.1).

Let's consider further from one variety given on the scheme the most made types of terrorist attacks in aircraft—explosion and taking of hostages.

(1) Explosion. Terrorists can put in action the explosive device as directly being in the place of commission of terrorist attack (suicide terrorism, the performer is the suicide bomber), and remotely, at the same time, the signal for activation of the explosive device can be transmitted as remotely (wire networks, radio a signal), and automatically, at achievement of the set conditions (the set time, these achievements of the set VS height and so forth). Explosive devices, as a rule,

Table 1.1 Information on the most considerable aviation incidents connected with terrorism during 2000–2019 [53–63]

№	Event	Number of dead/wounded (people)	Organizer of an event (terrorist attack)	Date of an event	Cause of the accident
1	Explosion in Airbus 321 air, flight No. 9268 (The Sinai Peninsula, Egypt)	224/0	Terr. org. "Vilait Sinai" (Islamic State)*	31.10.2015	Shortcomings of holding customs procedures (a bomb in a luggage compartment)
2	Flight No. MH17 (The Donetsk Region, Ukraine) is brought down in Boeing 777 air	298/0	Investigation is not completed	17.07.2014	It is brought down in air by a rocket of a series 9M38 of ZRK Buk/Is considered as terrorist attack
3	Hall of inter@ of arrivals of Domodedovo Airport (Moscow, Russia)	37/172	Terr. org. "Caucasus Emirate"*	24.01.2011	Shortcomings of holding customs procedures (penetration of the suicide bomber with a bomb in secur. area)
4	Explosion in air Tu-134A-3, flight No. 1303 (The Tula Region, Russia)	44/0	Terr. org. "The Karachay jamia"*	24.08.2004	Shortcomings of holding customs procedures, preflight control (penetration of suicide bombers with bombs on airplane)
5	Explosion in air Tu-154B-2, flight No. 1047 (The Rostov Region, Russia)	46/0			
6	Flight No. SBI 1812 is brought down in Tu-154 M air (232 km. from Sochi, Russia)	78/0	According to the Russian Federation report—it is brought down by air defense of Ukraine/officially it is not recognized	04.10.2001	It is brought down in air by a rocket of a series 5B28 of ZRK S-200 V/Other versions, including terrorist attack—explosion of a bomb onboard airplane

(continued)

Table 1.1 (continued)

№	Event	Number of dead/wounded (people)	Organizer of an event (terrorist attack)	Date of an event	Cause of the accident
7	Capture and use as the tool of terrorist attack Boeing 767-223ER, flight No. 11 (level 94–98 of floors VTTs-1, New York, USA)	2977/more 3000	Terr. org. Al-Qaeda*	11.09.2001	Complex shortcomings of ensuring anti-terrorist safety: holding customs procedures, preflight control and interpretation is more whole than passengers (a profayling, verification of documents and so forth), shortcomings of security of a cabin of crew of the aircraft from unauthorized access, lack of the reliable mechanism of tracking for stolen vozd. vessels, lack of means of air defense for the elimination of dangerous threats in air (shortcomings in administrative character on expeditious decision-making on the elimination of threats of means of air defense)
8	Capture and use as the tool of terrorist attack Boeing 767–222, flight No. 175 (level 78–85 of floors VTTs-2, New York, USA)				
9	Capture and use as the tool of terrorist attack Boeing 757–223, flight No. 77 (headquarters of the Ministry of Defence of the USA (Pentagon building), Arlington, Virginia, USA)				
10	Capture and use as the tool of terrorist attack Boeing 757–222, flight No. 93 (3 km. from Shanksvill, Pennsylvania, the USA)				

are self-made, are manufactured of the chemical elements and materials which are available on free sale, the charge pays off proceeding from specifics of an object of terrorist attack, with a necessary stock for the guaranteed achievement of "result". Explosion bears destructive, deadly consequences and if it occurred onboard the aircraft—practically in all cases to inevitable death of all persons which are on it at the crash. Terrorist attacks on the earth, in places of a congestion

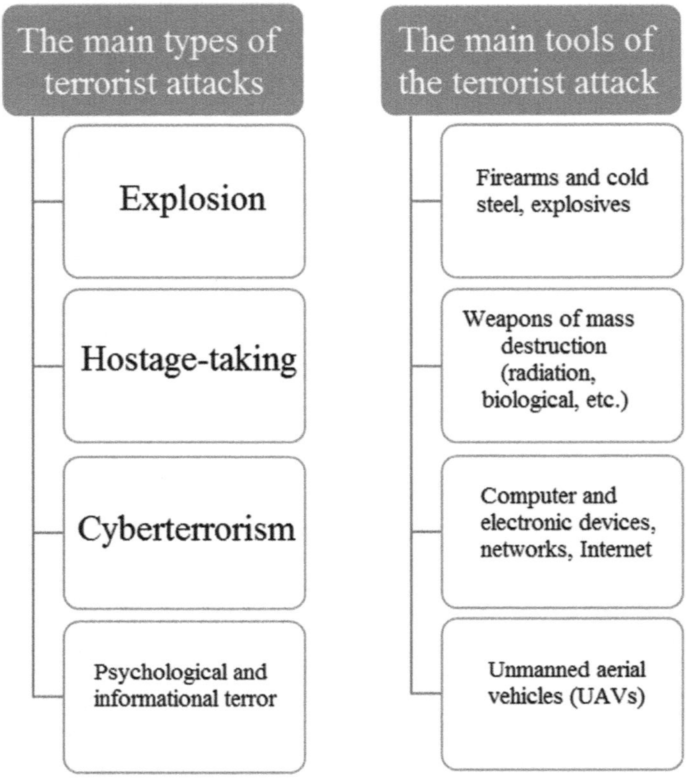

Scheme 1.1 Main types and instruments of commission of terrorist attacks in aircraft

of people on air terminal—also are extremely destructive and claim hundreds of the lives what events of the last tragedies described at the beginning of the research testify to. Explosion can be realized also by means of hit of a rocket or firing of the aircraft and also by collision with it of other object, nevertheless—similar scenarios of succession of events are outside the real research.

(2) Hijacking (stealing) of the aircraft with passengers onboard as one of types of terrorism is simpler in realization for terrorists, than explosion commission, however, and chances of success are much lower. The concept which settled in our country capture and stealing of the aircraft, recently also often call in the world the term "skyjacking" (English term) that means—capture, stealing of the aircraft, colloquial—the plane [42]. Hijacking of the aircraft can have the following tasks for terrorists.

(1) use of the aircraft—as weapon, as a subject of commission of more considerable terrorist attack (the most known example of similar actions was set by events on September 11, 2001 in the USA);

(2) capture in hostages of passengers and crew of the aircraft for the requirement from the authorities of performance of certain conditions (to release accomplices of terrorists from prison, to pay money and so forth).

Similar terrorist attacks can be conducted at the minimum use by terrorists of the forbidden objects and substances for the purpose of free pass aboard. The considerable number of hijacking of aircrafts was made in the world during the 1970–1990th years, at the same time, terrorist attacks were connected as with a political situation in the country (that was the phenomenon characteristic for those times in the USSR), and personal motives of aggressors, extremely low level of aviation safety (and in certain cases—almost total absence of security measures) —strongly promoted similar cases of manifestation of aggression.

Universal use of the computer equipment influences the formation of a new round of terrorist threat—computer terrorism (cyberterrorism). Development local (in airline) and global networks—generates (Internet), increase in a role of computers in navigation management and ensuring activity of airline itself and represents for terrorists new weak spots and the purposes for attack. At the same time, methods of impact on the computer equipment from terrorists can be various, for example:

(1) change (substitution) of these navigation devices (coordinates) in the equipment of the airport, the aircraft, and perhaps and directly via the satellite;
(2) impact on radio exchange between the aircraft and the air traffic controller (targeting of hindrances, substitution of signals);
(3) impact on the systems of protection, the automated control and management of access of airline for the purpose of creating favorable conditions for unauthorized access for terrorists to the territory, without holding customs procedures, bypassing an established order (at the same time, you should not forget that computer equipment is also the tool used and for different forgery of throughput documents which can be in turn used by terrorists);
(4) disorganization of work of airline, air terminal, by change of computer data on departures and arrivals of aircrafts, for the purpose of artificial increase in a mass congestion of people in insufficiently safe places (for the organization of further explosion, taking of hostages).

It is separately necessary to note a possibility of overcoming perimeter of the protected object of airline out of the established entrance zone with use of unmanned aerial vehicles (UAVs), which can also serve as the active instrument of commission of terrorist attack. Similar small aircraft are able how to bear explosive and toxic agents on the board, and by means of placed onboard transmitters to direct a radio noise at the equipment of the aircraft and a conference circuit that can entail catastrophic consequences. Now, for detection of the UAV widely use radar station, optical and acoustic sensors, means of interception of radio signals of information transfer, however, the special complexity is caused by detection of small aircraft [70]. In this regard, at terrorists small aircraft used to entertaining use and also self-made designs on their base at the heart of which completions, the possibility of increase in loading capacity for attachment of explosive and devices of destructive influence lies were widely used. For counteraction to similar devices, it is used both conventional classical firearms, and means of suppression and interception of radio signals and also the last achievements in the field of nonlethal weapon—the microwave oven and laser. The use of professional military UAVs did not gain the distribution in activity of terrorists, in connection with their high cost and complexity of production yet, the weight and dimensions, nevertheless, it is impossible to exclude that in the future similar aircraft will become more popular also with malefactors.

1.2 Standard and Legal Regulation of Anti-terrorist Safety

Acting as the element making impact on the system of anti-terrorist safety of airline, standard and legal regulation demands from airline of strict compliance with the law. A particular interest in the matter is that questions of legal support of anti-terrorist safety of airline in aspect of the real research, it is necessary to consider in four adjacent, but complementary standard and legal documents can be divided conditionally according to the following contents:

(1) questions of anti-terrorist safety;
(2) questions of aviation;
(3) questions of transport safety;
(4) standard and legal documents in which counteraction to terrorism, aviation and transport safety is considered generally, at one level.

Let's consider legislation questions on the example of the Russian Federation and the international normative legal acts. Legal bases of counteraction to terrorism in the Russian Federation are: constitution [1], international legal documents and rules of law, interstate contracts, federal laws, normative legal acts of executive authorities, and industry documents.

Consideration of a legal basis of fight against terrorism should be begun with the international standard and legal documents. For the first time, the question of counteraction to terrorism was raised in the European convention on suppression of terrorism of January 27, 1977 adopted to Strasbourg [5]. This convention adopted by the European countries defined need of acceptance of the effective measures aimed at providing inevitability of criminal prosecution of the persons involved in commission of acts of terrorism. Further continuation of this document was the Convention of the Council of Europe on prevention of terrorism adopted in Warsaw on May 16, 2005 [13], which, in difference from previous, most fully opens a package of measures for terrorism counteraction.

The Russian Federation (USSR) also accepted and became the participant of the following conventions of various levels.

- International convention on fight against taking of hostages (New York, on December 17, 1979) [6].

- Convention on physical protection of nuclear material (Vienna, on October 26, 1979) [7].
- The declaration on measures for elimination of the international terrorism (it is approved by the resolution of the United Nations General Assembly 49/60 of December 9, 1994) [8].
- International convention on fight against bombing terrorism (New York, on December 15, 1997) [9].
- The agreement on cooperation in the field of identification and overlapping of channels of penetration in the territory of member states of the Shanghai Cooperation Organization of the persons involved in terrorist, separatist and extremist activity (Shanghai, on June 15, 2006) [15].
- The contract on cooperation of the State Parties of the Commonwealth of Independent States in fight against terrorism (Minsk, on June 4, 1999) [10].
- The international convention on fight against terrorism financing (it is accepted by the resolution 54/109 of the United Nations General Assembly of December 9, 1999) [11].
- Shanghai convention on fight against terrorism, separatism and extremism (Shanghai, on June 15, 2001) [12].
- International convention on suppression of acts of nuclear terrorism (New York, on September 14, 2005) [14].
- Contract of the State Parties of the Commonwealth of Independent States on counteraction of legalization (washing) of criminal income and to terrorism financing (Dushanbe, on October 5, 2007) [16].
- The convention of the Shanghai Cooperation Organization against terrorism (Yekaterinburg, on June 16, 2009) [17].

As for the modern Russian legislation, in 2000, the Concept of national security of the Russian Federation in which terrorism was defined by legislators as real threat for sustainable and safe development of the country [25], at the same time, according to the author was accepted, the similar threat should be designated and much earlier, for example, in earlier existing Law of the Russian Federation "About safety" of 05.03.1992 No. 2446-1 [20].

For the first time, in article 5 of the Federal law of July 25, 2002 "About counteraction of extremist activity" preventive, information and propaganda measures, including educational character, on overcoming terrorism and extremism are established to No. 114-FZ [22].

In the Federal law of March 6, 2006 No. 35-FZ "About Counteraction to terrorism", definition is given to terrorism as "the ideological system of violence and practice of violent acts". Under counteraction to terrorism, the law considers activities of public authorities and local government for prevention of terrorism, including for identification and the subsequent elimination of the reasons and conditions promoting distribution of terrorist practice and ideology [23].

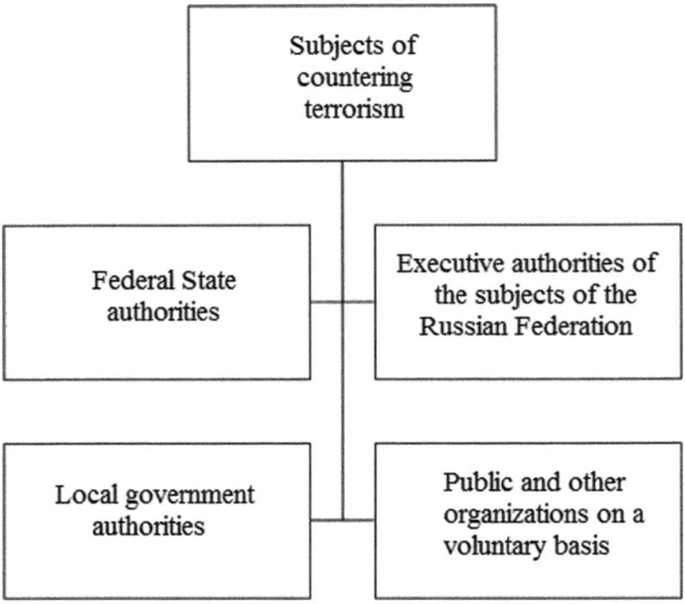

Scheme 1.2 Subjects of counteraction to terrorism [26]

In "The concept of counteraction to terrorism in the Russian Federation", approved as the Russian President of 05.10.2009 concrete preventive measures for the prevention of terrorism specified and the list of subjects of counteraction to terrorism is defined [26].

Subjects of counteraction to terrorism are (Scheme 1.2).

Concretizing the above-stated scheme, it is possible to allocate the following authorities and the organizations resisting to terrorism (Scheme 1.3).

Further development of the anti-terrorist legislative base is the improvement of Strategy of national security of the Russian Federation, which was approved by the decree of the Russian President of December 31, 2015 No. 683. In recent years, this document underwent multiple changes, in connection with dynamically changing world situation in the field of fight against terrorism. The concept of national security of the Russian Federation is understood as security of an individual, society and state from threats outside and from within at which realization of constitutional rights and freedoms of citizens of the Russian Federation, worthy quality and level of their life, sovereignty, independence, the state and territorial integrity, sustainable social and economic development of the Russian Federation is provided [27]. National security includes the defense of the country and all types of safety provided by the Constitution of the Russian Federation and the legislation of the Russian Federation, first of all, the state, public, information, ecological, economic, transport, energy security, safety of the personality [27].

Point 43 of Strategy distinguishes from the main threats of the state and public security, in relation to a subject of the real research the activity of the terrorist and

Scheme 1.3 Subjects of counteraction to terrorism (the concretized list) [26]

extremist organizations directed to violent change of the constitutional system of the Russian Federation, destabilization of work of public authorities, destruction or violation of functioning of military and industrial facilities, critical public infrastructure, transport infrastructure, intimidation of the population, including by taking with weapons of mass destruction, radioactive, poisoning, toxic, chemically and biologically dangerous substances, commission of acts of nuclear terrorism, violation of safety and stability of functioning of critical information infrastructure of the Russian Federation, the criminal encroachments directed against the personality, property, the government, public and economic security and also the natural disasters, accidents and accidents including connected with global climate change, deterioration in technical condition of infrastructure facilities and emergence of the fires [27].

In point 47 of Strategy, the concrete measures directed to increase in level of public and state security by prevention of their emergence on timely identification and prevention of threats are given [27].

The analysis of Strategy allows to draw a conclusion—terrorism is the main threat of national security of the Russian Federation. As the major object demanding safety transport infrastructure of the country is legislatively allocated. Thus, the state adequately estimates reality of threats from the terrorist organizations, and the frequency of tragic cases and the number of victims testify to need of increase in attention to safety on objects of transport infrastructure of the country.

Separate consideration is demanded by the international standard and legal documents and local documents of the Russian Federation devoted to directly aviation safety and fight against terrorism in aircraft to which numerous works of scientists-lawyers in the field of aircraft—A. Yu. Pidzhakova, G. A. Volkova, and others [43–46]. Not prevalence of civil aviation till 1950th years in the USSR positively affected also safety issues that can be noticed on lack of considerable incidents on air transport. With growth of number of air transportation and increase in availability of aircraft to citizens, also the first serious problems concerning safety in civil aviation began—numerous stealings of aircrafts and takings of hostages in the USSR considerably became frequent. The international conventions existing at that time (The convention on regulation of air movements (it is accepted in Paris 13.10.1919) [38], the Convention on the international civil aviation (it is accepted in Chicago 07.12.1944) [39] did not raise the questions of ensuring aviation safety and to counteractions to terrorist threats, in connection with almost total absence of precedents in this sphere. Besides, even, if the specified questions could be provided by conventions 1919 and 1944 of—doubtfully that for the USSR it would matter, in confirmation these words of the author need to address history—the USSR refused to participate and adopt the Chicago convention, the main reason for refusal as the Times newspaper of 30.10.1944 wrote was (quote): "… the invitation to this international forum of Spain, Portugal and Switzerland—the states which pursued for a number of years hostile policy concerning the Soviet Union" [51]. As a result, the Soviet Union joined the Chicago convention only in 1970. Similar "tactics" of ratification of the USSR of the international standard and legal documents mentioned also other the document, major for aircraft: The convention on the crimes and some other acts made onboard aircrafts (accepted in Tokyo, on September 14, 1963) which was adopted by Presidium of the Supreme Council of the USSR only on December 4, 1987 [2]. Nevertheless, other documents in the field of aviation safety were accepted by the USSR in due time: The convention on fight against illegal hijacking of aircrafts (adopted in the Hague, on December 16, 1970) is adopted by Presidium of the Supreme Council of the USSR on August 4, 1971 [3], the Convention on fight against the illegal acts directed against safety of civil aviation (adopted in Montreal, on September 23, 1971) is adopted by Presidium of the Supreme Council of the USSR on December 27, 1972 [4]. The provided international legal acts are directed to systematization of actions of the participating countries in questions of criminal prosecution of the persons responsible for illegal acts concerning civil aviation, a main objective is consolidation of efforts of all participating countries in ensuring aviation safety and fight against uniform threat. As dangerous acts are also allocated as acts of terrorism which objects are directly passengers VS and objects of transport infrastructure and also the terrorist attacks conducted with use of aircrafts as weapon for destruction of objects and people [41].

It is also necessary to pay attention and to attempts of the Russian Federation to bring the local legislation into accord to the recommendations of ICAO. Not absolutely successful first attempt of volume reduction of requirements of aviation safety in compliance with the recommendations of ICAO was made by development of "The program of aviation safety of civil aviation of the Russian Federation" on the basis of the order of the Ministry of Transport of the Russian Federation of 18.04.2008 No. 62 “About the approval of the Program of aviation safety of civil aviation of the Russian Federation” [32]. Nevertheless, this document was not actually entered into application, officially was not published in this connection the requirements listed in it could not be obligatory to execution that as a result was confirmed with the decision of the Supreme Court of the Russian Federation of September 11, 2019 No. AKPI19-529 [40], in this connection the Ministry of Transport of the Russian Federation the order of October 25, 2019 No. 344 repealed earlier issued order of the Ministry of Transport of the Russian Federation of 18.04.2008 No. 62 [32]. Also were adopted the order of the Ministry of Transport of the Russian Federation of March 5, 2011 No. 72 “About the approval of the Program of quality control of ensuring aviation safety of civil aviation of the Russian Federation” and the order of the Ministry of Transport of the Russian Federation of March 5, 2011 No. 73 “About the approval of the Program of preparation in the field of aviation safety in civil aviation of the Russian Federation” which were canceled, nevertheless, already on June 2, 2011 by the relevant order of the Ministry of Transport of the Russian Federation No. 153.

Considering the Russian normative legal acts in the sphere of ensuring aviation safety, it is necessary to address, first of all, the profile code—“The air code of the Russian Federation” of 19.03.1997 No. 60-FZ that separately does not allocate terrorism, but, puts the general sense as illegal act or inaction in Art. 83 of the code devoted to aviation safety. Within this book, anti-terrorist safety is considered as a concept of structure aviation safety which is defined by the code as “a condition of security of aircraft from illegal intervention in activity in the field of aircraft” [19], whereas act of terrorism is one of types of illegal intervention in activity in the field of aircraft.

The federal law “About Transport Safety” of 09.02.2007 No. 16-FZ defines main objectives and problems of ensuring transport safety in the Russian Federation. The concept “transport safety” is considered as a complex of certain measures in the field of ensuring transport safety and as subject to legal regulation [24]. The purposes of ensuring transport safety are: ensuring steady and safe functioning of a transport complex, protection of interests of the personality, society and state in the sphere of a transport complex from acts of illegal intervention, at the same time is specified a number of the main objectives and nodal principles of ensuring transport safety [24].

Part 1 of article 4 of the law provides to assign ensuring transport safety of objects of transport infrastructure and vehicles to subjects of transport infrastructure, carriers if other is not established by the law and other federal laws while a part 2 laws says that federal executive authorities ensure transport safety according to the competence established by the specified law, other federal laws and other regulations of the Russian Federation adopted according to them. Executive authorities of territorial subjects of the Russian Federation, local governments participate in ensuring transport safety according to the competence established by the present Federal law, other federal laws and other regulations of the Russian Federation adopted according to them. Paragraph 6 of article 4 of the above-stated law also ranks acts of terrorism as actions that can be acts of illegal intervention in activity of transport infrastructure. Point 5 of part 1 of article 9 of the law, specifies that for planning and implementation of measures for ensuring transport safety, the persons brought in the list of the organizations and natural persons concerning whom there are data on their participation in extremist activity or terrorism, according to the Federal law of August 7, 2001 No. 115-FZ "About counteraction of legalization (washing) of income gained in the criminal way and to terrorism financing" have no right to be involved in the works which are directly connected with ensuring transport safety [21].

The state control and supervision in the field of ensuring transport safety are carried out by authorized federal executive authorities according to the legislation of the Russian Federation. According to the law, an indispensable condition of activity of an object of transport infrastructure is carrying out assessment procedure of its vulnerability. Article 6 of the law also provided categorization of an object of transport infrastructure, and article 7 regulates the establishment of levels of safety of objects of transport infrastructure and vehicles. The specified measures allow to systematize and classify objects by safety levels, to state regulate the procedure assessment of vulnerabilities on purpose increase in responsibility and to giving of due control of these actions from carriers.

The order of the Ministry of Transport of the Russian Federation of November 28, 2005 No. 142 "About the approval of the Federal aviation rules "Requirements of Aviation Safety to the Airports" in point 70 of the section VIII establishes that for prevention of acts of terrorism on objects of the airports, including at the station square and territories adjoining to it, on entrances to air terminal and in buildings of air terminals and also in zones of control of passengers and points of examination the events for prevention and control of offenses, protection of public order and ensuring public safety coordinated with Internal Affairs Service are held [31]. The normative legal act is the document containing an exhaustive list of the requirements allowing to provide the admissible level of safety on objects of aviation infrastructure.

In requirements of aviation safety to the airports concrete requirements to the organization and providing are described [31].

- aviation safety at the airport;
- the access and inside control at the airport;
- protection of aircrafts and objects of the airport;
- preflight and post-flight examination in the territory of the airport;
- preflight examination of aircrafts;
- actions for settlement of the emergency situations connected with acts of illegal intervention in activity of civil aviation;
- features of aviation safety at the airports of the Moscow airline hub, the international airports and the airports of federal importance.

The resolution of the Government of the Russian Federation of 10.12.2010 No. 940 "About levels of safety of objects of transport infrastructure and vehicles and about an order of their announcement (establishment)" [28], and also the resolution of the Government of the Russian Federation of 28.07.2018 No. 886 "About the approval of requirements for ensuring transport safety, including requirements to anti-terrorist security of objects (territories) considering safety levels for various categories of objects of transport infrastructure and vehicles of air transport" [29], and also the order of the Ministry of Transport of the Russian Federation of 12.04.2010 No. 87 "About an order of evaluating vulnerability of objects of transport infrastructure and vehicles" [33] oblige the air enterprises to carry out assessment of vulnerability. The order of the Ministry of Transport of the Russian Federation of 21.02.2011 No. 62 defines an order of categorization of objects of transport infrastructure to this or that category, depending on characteristics and features of an object [34]. The main objective of categorization of similar objects is the assessment of their vulnerability for preliminary identification of shortcomings and objectively weak points of the system of providing measures of aviation safety, for acceptance of immediate measures for elimination of these shortcomings. Categorization is made taking into account the degree of threat to possible illegal intervention in a transport object field of activity, at the same time, exclusively accredited organizations authorized for these actions and entering the corresponding register of the accredited organizations can carry out similar assessment. Results of assessment together with the plan of safety of an object of transport infrastructure are without fail shown in Federal Air Transport Agency for their statement.

According to an order of the Government of the Russian Federation in 2002 in the Ministry of Industry and Science of Russia the methodical documents concerning questions of categorization and assessment of safety are developed [35, 36].

- "Methodical recommendations about categorization of objects of science, industry, power and life support about degree of their potential danger and subversive and terrorist vulnerability".
- "Methodical recommendations about assessment of sufficiency of actions for physical protection and protection of the enterprises and the organizations from the existing and predicted diversionary and terrorist threats of internal and external character".

The provided recommendations allow to systematize knowledge and to order actions by assessment and identification of shortcomings of the system of ensuring security of objects from subversive and terrorist threats. Methodical recommendations have the general character and extend to all types of objects of transport, including aviation.

A specific place among the Russian standard and legal documents of the present in the field of aviation safety is held by the Federal system of ensuring aviation safety (The national program of aviation safety) developed according to the guide of ICAO to aviation safety and approved by the Interdepartmental commission on aviation safety, safety of flights of civil aviation and simplification of formalities [30]. The specified document represents the list of specific actions for operators of airlines for ensuring aviation safety and safety of flights, at the same time the document is deprived of an excessive formalism (as well as it is noted in the name of the commission which approved it), with the indication of both the general tasks, and quite concrete—up to settings of sensitivity of the customs equipment and requirements to the personnel involved for similar activity. Authors of Federal system of ensuring aviation safety could state in 110 pages of the text the main contents of standard and legal base in the field of aviation safety and safety of flights, complemented with specific proposals and features for this kind of activity, brought documentation into accord with an all-standard type of the documents applied by ICAO.

Disclosing the main objectives of a research, the term "system of anti-terrorist safety of the air enterprise" needs to be adapted and treated as a package of measures, including human and technical resources, intended for protection of the air enterprise against terrorist influence. Legally, the concept—"anti-terrorist safety of the air enterprise" should be treated as a condition of security of the air enterprise from terrorist influence in activity of airline.

The carried-out analysis of normative legal acts and auxiliary documents allows drawing a conclusion that in connection with increase of activity of the terrorist organizations, the Russian Federation takes the necessary measures directed to consolidation and the general systematization of knowledge, methods, and forms of fight directed to counteraction to terrorism the corresponding effective rule-making work conducted. The existing standard and legal documentation accurately classifies terrorism as the main threat of national security of the Russian Federation, and the local methodical recommendations about categorization of objects and assessment of vulnerability have to allow revealing in due time objects and their components, the most subject to the attack of terrorists. If regarding the legal base of counteraction to terrorism, it is possible to draw a conclusion on the existence of sufficient and full data and the carried-out work, then questions of distribution of a financial resource, the choice of the most correct techniques of strengthening of technical base and selection of shots are objectively insufficiently opened, at the same time, it is necessary to draw an intermediate conclusion that terrorism on objects of aviation infrastructure still is extremely relevant what the frequency of incidents and the attention paid to this perspective from the state testifies to.

1.3 Technical and Economic Features of Counteraction to Terrorism

In the analysis of anti-terrorist safety as one of the making aviation safety influencing the general organization of activity of airline without failures expedient is to study technical and economic features of counteraction of terrorist activity. Studying of economic features of the terrorist organization, first of all, is necessary for understanding of mechanisms of functioning of the terrorist organization, assessment the cost of preparation and commission of act of terrorism, i.e. "how many", directly, manage to organize and conduct a terrorist attack for terrorists (Schemes 1.4 and 1.5).

At the same time, it is admissible to make a hypothesis—the low cost of preparation of a terrorist attack will surely lead to growth of their number and will increase risk of their successful commission for terrorists whereas the cost of elimination (elimination) of consequences of terrorist attack (payments to the victims and victims,

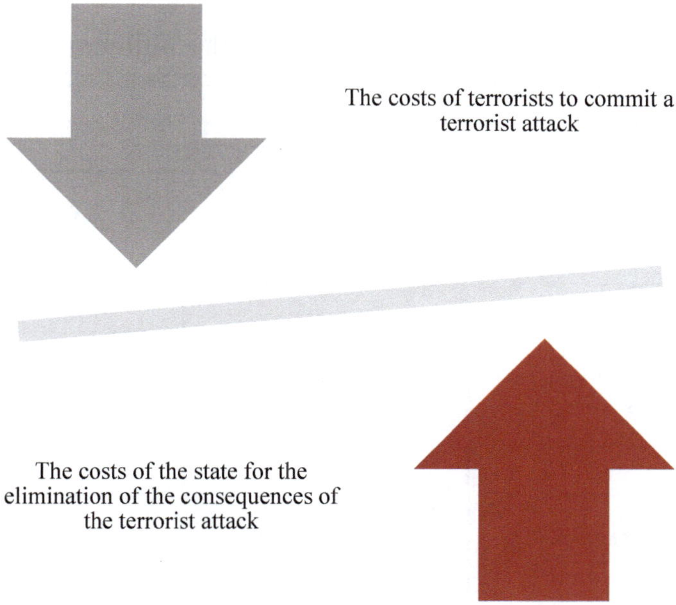

Scheme 1.4 Expenses of the resisting parties

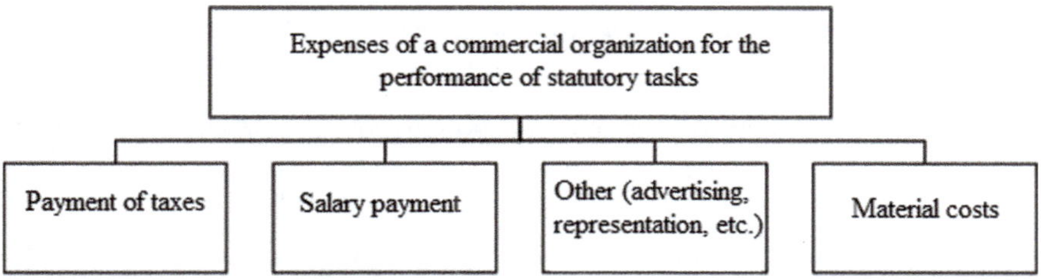

Scheme 1.5 Costs of the commercial organization of performance of authorized tasks

restoration of infrastructure of airline, aircrafts and so forth) are extremely expensive and do not come within miles of low expenses of terrorists (Scheme 1.6).

For assessment of costs of terrorists of preparation and commission of terrorist attack, it is necessary to consider functioning of the terrorist organization (or the individual terrorist)—as work of the usual commercial organization or the individual entrepreneur. At it, quite sample, terrorism categorization—as one of the types of commercial activity, it turns out most clear, is evident and proved to estimate all expenses connected with performance of the main authorized objective of such enterprise—commission of act of terrorism with the smallest expenses (expenses) and causing the maximum harm for the world around (destructiveness).

Costs of any organization of performance of the authorized tasks can be structured by following a look (Scheme 1.5).

Scheme 1.6 Costs of the terrorist organization of performance of an authorized task—terrorist attack commission

At the same time, it should be noted that the terrorist organization cannot be officially registered, have certain obligatory expenses (contributions to funds, payment of taxes and so forth) in connection with the obviously illegal activity forbidden practically in all civilized countries of the world. Undoubtedly, there are terrorist organizations that mask the true authorized tasks under law-abiding, and conducting usual economic activity is carried out by them only for cover and legalization of illegal acts, however, considering direct sources of income and expenses on commission of acts of terrorism, it is necessary to analyze their functioning only in the simplified option. Is expedient to revise the similar scheme according to its applicability directly to a terrorist image of activity. At the same time, can be sources of financing of the terrorist organization and as the activity bringing income from legal business, and the money earned in the criminal way—commission of other types of crimes (except terrorism), reception of money from the third parties—donations, on financing of terrorist actions. Nevertheless, the general budget of the terrorist organization consists as of the money received in the criminal way and the means, which are initially not received in the criminal way that does not matter for achievement of final authorized tasks of the organization. The bank organizations and executive authorities are active on counteraction of money laundering, received in the criminal way—active work on blocking of operations when carrying out doubtful bank transfers is for this purpose conducted, additional data from senders and recipients of money for confirmation of a source of origin of such means are requested. Despite the applied measures, the maintenance of legal commercial structures by terrorists and various tricks on a legend (conspiracy) service of the activity allow to bypass the standardized automatic blocking of accounts and suspicious operations, and the presence of corrupt officials and the workers sympathizing terrorists in the bank sphere, does similar control measures for sides of the state, the practical terrorists deprived of sense and directed only to control for insufficiently skilled and qualified which quantity steadily decreases. Despite it, improvement of similar control measures behind the financial sector of shadow economy has led to decrease in an opportunity quick to redirection of cash flows for financing of terrorism by means of non-cash ways in the near future, however, will inevitably lead to other ways of movement of money—in a cash form. Transition to a cash form of calculations is a priority for most of shadow financial agents also at present, and in a case with control strengthening—will become the compelled and inevitable measure, at the same time, it is necessary to believe that there will be also a relevant task of physical movement of considerable money between the cities and the countries.

Costs of the organization of criminal activity are conditionally shown on Scheme 1.6.

The represented scheme demonstrates that activity of the terrorist organization is out of the legal framework in a type of the authorized task, and on the general structure and expenses (collecting intelligence information, expenses on conspiracy and a legend-methodic and so forth)—is similar to the staff of the state, which are engaged operational search and intelligence activities. The lack of need of payment of taxes and contributions to funds, in turn, has to be opposed to the increased expenses on secret and current ensuring activity as premature disclosure of the terrorist organization in connection with insufficient safety can entail the termination of its functioning and be a failure of performance of the main authorized objective.

Studying questions of an account part of the terrorist organizations, it is necessary to consider in detail sources of financing of terrorist activity (Scheme 1.7).

It is impossible to claim unambiguously what will be the estimated main source of financing of terrorist communities. In the budget of the terrorist organization sponsor's financing from individuals and the commercial organizations, sympathizing terrorist activity can make the impressive sum. At the same time, rendering the similar sponsor's help, these persons and the organizations can pursue own aims at which the terrorist organization becomes in their hands the tool for the embodiment of concrete criminal plans. Similar interpretation of activity concerns also representatives of the hostile states, which are ready to spend a considerable financial resource for strengthening of terrorists for the purpose of destabilization of a political and/or economic situation in a certain country to please of the interests.

One of the purposes of the present paragraph is to estimate and compare expenses of terrorists directly on commission of acts of terrorism with their destructive consequences for the state and society, considering open (not confidential) character of the

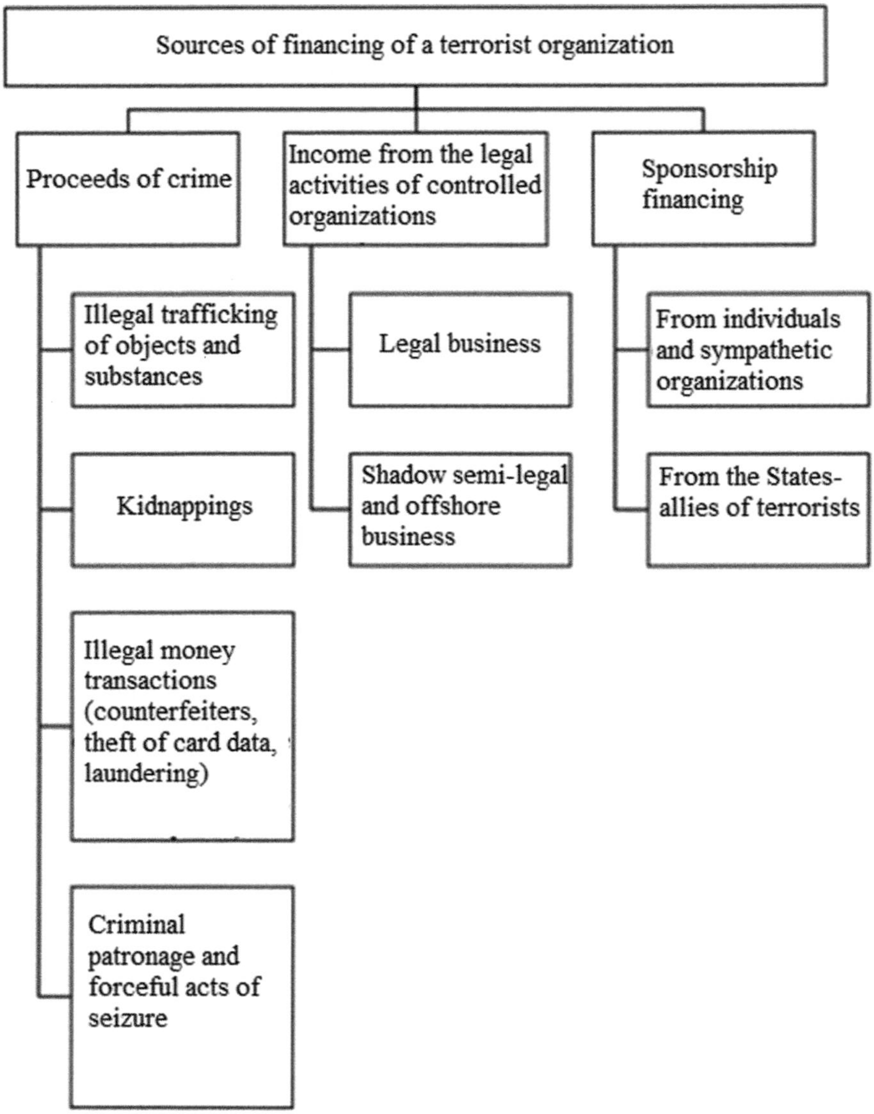

Scheme 1.7 Sources of financing of the terrorist organization

real research. Further the available data on damage caused by terrorist attacks and also data on the cost of expenses on their commission are considered.

The most extensive damage for all history of mankind is caused by terrorist attacks in the USA on September 11, 2001. The terrorist attacks conducted by members of the Al-Qaeda terrorist organization * by means of hijacking of aircrafts claimed the lives not only of their passengers but also the people who are in World Trade Center and in the Pentagon building. The general expenses of the U.S. Government on mitigation of consequences of terrorist attack made about $3.3 trillion. This sum develops from: expenses on physical damage and restoration, expenses on modernization of security systems, expenses connected with the termination of business process and reduction of tourist flow, increase in military financing in connection with the taken place events, assignment for future war and the help to veterans of war [52] At the same time, the cost (expenses) of the international terrorist organization "Al-Qaeda"* on the organization and commission of terrorist attacks was made by only about $600 thousand [53].

On January 24, 2011 act of terrorism happened in the hall of the international arrivals of the Moscow Domodedovo Airport. The explosion made by the suicide bomber claimed the lives of 37 people, at the same time, 172 were wounded [54]. There are data on total amounts of payments to victims and the dead in the range from 150 to 300 million rubles ($4.9 million to $9.8 million), but the amounts of damage, put to airline—are not called [55]. Terrorist attack cost the Caucasus Emirate terrorist organization* in the sum about 500 thousand rubles ($16,000) [56].

On April 3, 2017 in St. Petersburg (Russia) on a stage between the Sennaya Square and Institute of technology stations, in the car of the train of the subway, the suicide bomber carried out blasting the improvised explosive device. The second planned terrorist attack which I am assumed to commit blasting a bomb with remote control at the Vosstaniya Square metro station managed to be prevented thanks to vigilance of personnel of the subway and field services. It is known of 103 victims and 16 dead. Payments of the state to relatives of the dead and the victim made in the total amount of 30.5 million rubles [65]. The cost of the state's expenses for liquidation and preventive measures related to the terrorist attack was not officially reported. If we proceed from the available information on costs, then the State Unitary Enterprise "St. Petersburg Metro" in 2008 purchased similar cars worth 18 million rubles each [66]. Respectively, data on damage can be made of the cost of one destroyed car of intermediate type and payments to dead and injured of 30,5 rubles that in total amount makes 48,5 million rubles ($866 thousand). According to official data of Rosfinmonitoring, this terrorist attack cost his organizers of all 150 thousand rubles ($2680) [67].

On December 27, 2002, the leader of terrorists Shamil Basayev by means of the Chechen fighters organized terrorist attack at the government house of Chechnya. Act of terrorism was carried out by means of blasting the KAMAZ and UAZ trucks by suicide bombers by which they got into a complex of government buildings. Total number of the victims was 71 persons, also various degrees of severity 640 people were wounded. The damage from terrorist attack was 267.7 million rubles ($8.4 million), including cost destroyed at explosion of buildings and property and funds, allocated for payments of compensations by the victim and to families of the dead. The cost of costs of terrorists of preparation and commission of terrorist attack was only about 570 thousand rubles ($18,000) [68].

On April 24, 1993 in Bishopsgeyt (London, England), terrorists of the Irish Republican Army (IRA)* conducted a terrorist attack by explosion of the dump truck filled with explosive. Because events took place in the morning on Saturday, in a quarter where there were practically no residential buildings, the number of victims was succeeded to avoid (there is information about 44 wounded and 1 dead). Nevertheless, this terrorist attack can also be ranked as one of the most impressive on the caused damage—about $1.2 billion. Costs of terrorists of IRA of preparation and commission of terrorist attack made only $4800, besides they even managed to save at the organization of terrorist attack—for commission of terrorist attack the dump truck which is earlier hijacked and recolored in other color was used [69, 71].

The carried-out review of data on acts of terrorism, expenses on their commission from terrorists and also expenses of the states on payments and mitigation of consequences, demonstrates that even at insignificant financing of terrorism—the damage in hundreds, and even thousands of times surpasses the cost of the organization of terrorist attacks.

Considering the general growth of number of air transportation and forecasts for increase in a passenger traffic and the transported freights for the next decades [49], it is necessary to believe that this progress will surely cause increase in appeal of objects of civil aviation as targets for terrorism. Under the specified circumstances, act of terrorism has to be prevented at any cost, security measures on objects, including aviation, transport infrastructure, are obliged to have character of the super reliable, minimizing opportunities both casual, and deliberate human errors and the equipment.

Task, widespread in the equipment and economy, is the analysis of a ratio of expenses and the achieved result. It is logical to assume that by what more financial resource is spent for realization of certain actions, especially the final result (linear dependence) will be essential, besides it is obvious—without any expenses or in case of their extreme insignificance—the result will be low or will not be reached. In a case with terrorism and expenses as on execution of terrorist attacks, and means for mitigation of consequences, we will assume the return hypothesis—costs of terrorists of the organization of terrorist attack do not predetermine causing bigger damage to the state and citizens and also that terrorist attack commission even without excessive expenses (for example, from ideological beliefs) cannot have under itself financings at all—as bases of carrying out destructive and destructive actions, at the same time the damage to the state and society anyway will be considerable.

For check of a hypothesis of lack of dependence between the financial resource spent by terrorists for the organization of terrorist attack and actually caused by damage it is necessary to compare data: (a) about costs of terrorists of the organization of terrorist attack; (b) about the actual damage caused by terrorists. It is convenient to represent the comparison of the available data on expenses of the terrorists and means spent for mitigation of consequences in the form of schedules on which it is evident to provide information.

Let's consider data on costs of terrorists of the organization and commission of terrorist attack at which expenses made from $2680 to $600 thousand (Schedule 1).

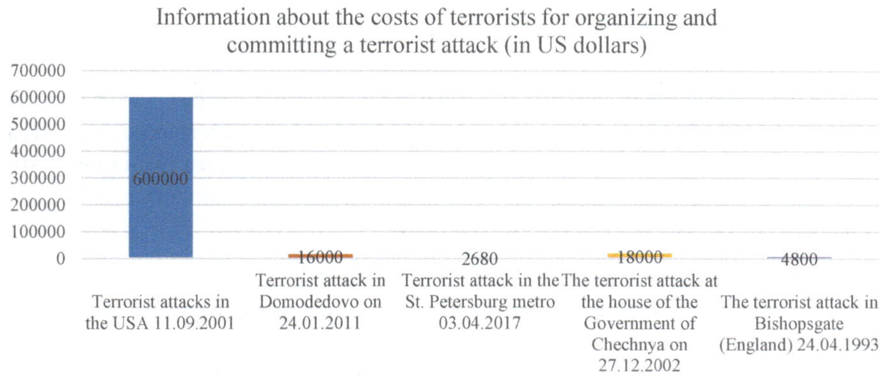

Schedule 1

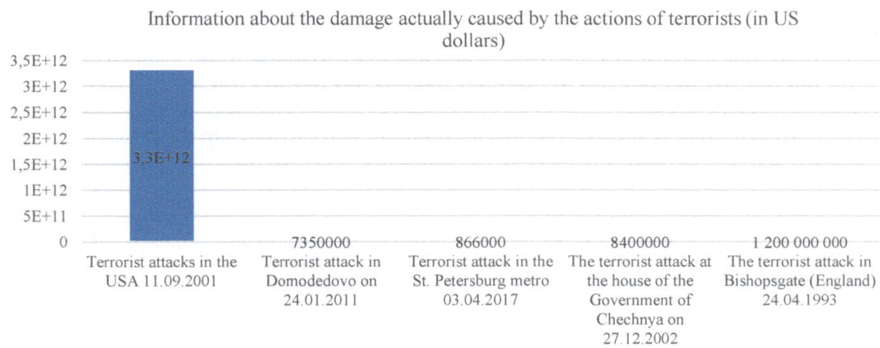

Schedule 2

Data on costs of terrorists of the organization and commission of terrorist attack, let's consider data on costs of the state (airline) of mitigation of consequences (Schedule 2).

The analysis and comparison of data on schedules allow to say with confidence that between costs of terrorists of the organization and commission of terrorist attack and damage (funds for mitigation of consequences), which is actually caused to the state that there is no dependence. It is also necessary to note that commission of acts of terrorism only from ideological and religious beliefs at which they are made with the minimum expenses is widespread. The carried-out analysis of lack of dependence allows to draw a conclusion that the scientific problem of decrease in risk of realization of terrorist threats in airline objectively is the most important in ensuring activity of airline—threat for life and health of passengers and multimillion costs of mitigation of consequences, even the terrorist layman with the minimum budget, as well as the fanatic acting only from ideological beliefs, practically without financial resource can create and embody in action.

1.4 Conclusion of the Chapter

1. The concepts existing in normative legal acts (including, nodal) concerning the organization of anti-terrorist activity can ambiguously be interpreted in relation to the questions that are directly connected with a terrorism problem in aircraft. It objectively demands the corresponding adaptation of certain terms, which are expedient for treating in the following look.

   - "*the system of anti-terrorist safety of the air enterprise*" defined as a package of measures, including human and technical resources, intended for the protection of the air enterprise against terrorist influence.
   - "*aviation safety*"—the condition of security of the air enterprise from acts of illegal intervention in their activity, including terrorist orientation providing acceptable risk level of such intervention.
   - "risk"—is considered as a combination of a possibility of approach (commission) of terrorist attack and the related adverse effects.
   - "vulnerability"—is the parameter characterizing a possibility of causing potential harm and damages to an object of aviation safety (infrastructure, personnel, passengers) by means of act of terrorism.

2. Regarding safety from terrorist threats it is necessary to carry to the main activities of airline:

   - standard legal support;
   - definition of threats of commission of acts of illegal intervention;
   - assessment of vulnerability and categorization of objects;
   - development and implementation of requirements for safety;
   - actions for settlement of the emergency situations connected with acts of illegal intervention in activity of civil aviation;
   - features of aviation safety at the airports of the Moscow airline hub, the international airports and the airports of federal importance.

3. It is necessary to carry to requirements of anti-terrorist safety of the airports:

   - throughput and inside the mode at the airport;
   - protection of aircrafts and objects of the airport;
   - preflight and postflight examination in the territory of the airport;
   - preflight examination of aircrafts.

   The specified requirements also define a task of ensuring anti-terrorist safety of airlines and set the relevant organizational structure.

4. The carried-out analysis showed that between costs of terrorists of the organization and commission of terrorist attack and the damage, which actually caused to the state and society that there is no dependence. On the basis of the data generalized in a research, it is necessary to draw a conclusion that the scientific problem of decrease in risk of realization of terrorist threats objectively is the most important in ensuring activity of airline in the field of aviation safety—threat for life

and health of passengers and multimillion costs of mitigation of consequences, even the terrorist layman with the minimum budget, as well as the fanatic acting only from ideological beliefs that is visually shown in the examples given in the head can create and embody in action.

5. It is expedient to structure technical means of system of ensuring anti-terrorist safety as follows: the television systems of observation, the security alarm system and the automated control and management systems for access, the equipment of examination and control, the system of fire safety, engineering protective constructions and warning systems. At the same time, the last boundary of prevention of a bringing of the forbidden objects and substances are customs procedures of passengers and baggage regarding existence/lack of the forbidden objects/substances. The analysis of technical means of system of anti-terrorist safety allows to consider technical means as the highly reliable *N-element* systems of a complex of anti-terrorist safety of airline, at the same time failure only one of them (!)—can lead to terrorist attack commission that needs to be considered further for the solution of a task on the distribution of resources.

References

1. The constitution of the Russian Federation (accepted by national vote 12.12.1993)//Reference and information system Consultant+
2. The convention on the crimes and some other acts made onboard aircrafts (Tokyo, on September 14, 1963)//Reference and information system Consultant+
3. Convention on fight against illegal hijacking of aircrafts (Hague, on December 16, 1970)//Reference and information system Consultant+
4. The convention on fight against the illegal acts directed against safety of civil aviation (Montreal, on September 23, 1971)//Reference and information system Consultant+
5. European convention on terrorism suppression (Strasbourg, on January 27, 1977) ETS N 090//Reference and information system Consultant+
6. International convention on fight against taking of hostages (New York, on December 17, 1979)//Reference and information system Consultant+
7. Convention on physical protection of nuclear material (Vienna, on October 26, 1979)//Reference and information system Consultant+
8. The declaration on measures for elimination of the international terrorism (it is approved by the resolution of the United Nations General Assembly 49/60 of December 9, 1994)//Reference and information system Consultant+
9. International convention on fight against bombing terrorism (New York, on December 15, 1997)//Reference and information system Consultant+
10. The contract on cooperation of the State Parties of the Commonwealth of Independent States in fight against terrorism (Minsk, on June 4, 1999)//Reference and information system Consultant+
11. The international convention on fight against terrorism financing (it is accepted by the resolution 54/109 of the United Nations General Assembly of December 9, 1999)//Reference and information system Consultant+
12. Shanghai convention on fight against terrorism, separatism and extremism (Shanghai, on June 15, 2001)//Reference and information system Consultant+

13. Convention of the Council of Europe on prevention of terrorism (Warsaw, on May 16, 2005)//Reference and information system Consultant+
14. International convention on suppression of acts of nuclear terrorism (New York, on September 14, 2005)//Reference and information system Consultant+
15. The agreement on cooperation in the field of identification and overlapping of channels of penetration in the territory of member states of the Shanghai Cooperation Organization of the persons involved in terrorist, separatist and extremist activity (Shanghai, on June 15, 2006)//Reference and information system Consultant+
16. Contract of the State Parties of the Commonwealth of Independent States on counteraction of legalization (washing) of criminal income and to terrorism financing (Dushanbe, on October 5, 2007)//Reference and information system Consultant+
17. The convention of the Shanghai Cooperation Organization against terrorism (Yekaterinburg, on June 16, 2009)//Reference and information system Consultant+
18. "Criminal Code of the Russian Federation" of 13.06.1996 No. 63-FZ//Reference and information system Consultant+
19. "The air code of the Russian Federation" of 19.03.1997 No. 60-FZ//Reference and information system Consultant+
20. The act of the Russian Federation "About safety" of 05.03.1992 No. 2446–1//Reference and information system Consultant+
21. Federal law "About Counteraction of Legalization of Income Gained in the Criminal Way and to Terrorism Financing" of 07.08.2001 No. 115-FZ//Reference and information system Consultant+
22. Federal law "About Counteraction of Extremist Activity" of 25.07.2002 No. 114-FZ//Reference and information system Consultant+
23. Federal law "About Counteraction to Terrorism" of 06.03.2006 No. 35-FZ//Reference and information system Consultant+
24. Federal law "About Transport Safety" of 09.02.2007 No. 16-FZ//Reference and information system Consultant+
25. The decree of the Russian President of 10.01.2000 No. 24 "About the Concept of national security of the Russian Federation"//Reference and information system Consultant+
26. "Concept of counteraction to terrorism in the Russian Federation" (accepted by Russian President 05.10.2009)//Reference and information system Consultant+
27. The decree of the Russian President of 31.12.2015 No. 683 "About the Strategy of national security of the Russian Federation"//Reference and information system Consultant+
28. The resolution of the Government of the Russian Federation of 10.12.2010 No. 940 "About levels of safety of objects of transport infrastructure and vehicles and about an order of their announcement (establishment) "//Reference and information system Consultant+
29. The resolution of the Government of the Russian Federation of 28.07.2018 No. 886 "About the approval of requirements for ensuring transport safety, including requirements to anti-terrorist security of objects (territories) considering safety levels for various categories of objects of transport infrastructure and vehicles of air transport"//Reference and information system Consultant+
30. "The federal system of ensuring aviation safety (The national program of aviation safety) " (approved by the Interdepartmental commission on aviation safety, safety of flights of civil aviation and simplification of formalities 15.10.2019)//Reference and information system Consultant+
31. The order of the Ministry of Transport of the Russian Federation of 28.11.2005 No. 142 "About the approval of the Federal aviation rules "Requirements of Aviation Safety to the Airports" (It is registered in the Ministry of Justice of the Russian Federation 28.12.2005 N 7321)//Reference and information system Consultant+
32. The order of the Ministry of Transport of the Russian Federation of 18.04.2008 No. 62 "About the approval of the Program of aviation safety of civil aviation of the Russian Federation"//Reference and information system Consultant+

33. The order of the Ministry of Transport of the Russian Federation of 12.04.2010 No. 87 “About an order of evaluating vulnerability of objects of transport infrastructure and vehicles”//Reference and information system Consultant+
34. The order of the Ministry of Transport of the Russian Federation of 21.02.2011 No. 62 “About the Order of establishment of amount of categories and criteria of categorization of objects of transport infrastructure and vehicles by competent authorities in the field of ensuring transport safety”//Reference and information system Consultant+
35. Methodical recommendations about categorization of objects of science, industry, power and life support about degree of their potential danger and subversive and terrorist vulnerability. Are approved by the Order of the Ministry of Industry and Science of Russia of 29.05.2002 N 145 (an order of the Government of the Russian Federation of 25.04.2002 of N HV-P4–06095)//Reference and information system Consultant+
36. Methodical recommendations about assessment of sufficiency of actions for physical protection and protection of the enterprises and the organizations from the existing and predicted diversionary and terrorist threats of internal and external character. Are approved by the Order of the Ministry of Industry and Science of Russia of 29.05.2002 N 145 (an order of the Government of the Russian Federation of 25.04.2002 of N HV-P4–06095)//Reference and information system Consultant+
37. Appendix 17. Safety. Protection of the international civil aviation against acts of illegal intervention. Prod. 9. Montreal: ICAO, 2011. 44 pages//Reference and information system Consultant+
38. Convention on regulation of air movements (Paris 13.10.1919)//Reference and information system Consultant+
39. Convention on the international civil aviation (Chicago 07.12.1944)//Reference and information system Consultant+
40. Decision of the Supreme Court of the Russian Federation of September 11, 2019 of N AKPI19–529//Reference and information system Consultant+
41. Abashidze AK (2019) Old, but modern problems of the international air law. Abashidze AK, Travnikov AI//Right. Magazine of Higher School of Economics. No. 3, pp 181–202
42. Baryshnikov SG (2011) Structure of a hijacking: social and philosophical analysis. Bulletin of the Perm national research polytechnic university. Soc Econ Sci (12):45–50
43. Volkov GA (2012) Organizational and legal bases of ensuring aviation safety. Wolves GA, Pidzhakov AYu, Shakhbazov RA (2012) Studies. Grant. SPb.: HECTARE university, 188 pp
44. Pidzhakov AY (2009) State regulation in the field of aviation safety: shortcomings and ways of improvement. Pidzhakov AY, Lebedeva MYuM (2009) Scientific bulletin of the Moscow state technical university of civil aviation (144):76–80
45. Pidzhakov AY (2009) The legal nature and problems of practical application of the international aviation regulations regulating questions of ensuring aviation safety. Pidzhakov AY, Malkova EGM (2009) Scientific bulletin of the Moscow state technical university of civil aviation (144):62–71
46. Pidzhakov AY (2013) Definition “transport safety”. Pidzhakov AY, Lebedeva MYuM (2013) Scientific bulletin of the Moscow state technical university of civil aviation (196):67–71
47. Hlobustov OM of KGB of the USSR 1954–1991 (2015) The mysteries of death of the Great power. Publishing Center “Akwa Term”, 656 pp
48. Horomin Nya (2007) Encyclopedia of wisdom. The collection of thoughts, sayings, aphorisms, paradoxes, epigrams. “Ripol Klassik”, p 482
49. Chinyuchin YM (2014) A general characteristic of the state of affairs in civil aviation and problematic issues. Scientific bulletin of the Moscow state technical university of civil aviation (205):5–9
50. Chernyadyeva NA (2013) Was terrorism in the ancient world? To a question of prerequisites and factors of formation of modern terrorism. Bulletin of the Perm university. Jurisprudence, No. 2(20), pp 32–39
51. Schenkman J (1955) International Civil Aviation Organization. Library E. Droz, Geneva, p 175

52. interfax.ru [URL]. Terrorist attacks cost on September 11 the USA to $3,3 trillion—a research. Interfax news agency. https://www.interfax.ru/russia/207362. Date of the address: 27.06.2019
53. centrasia.org [URL]. D. Sedov: American dream. Crime without punishment. Centraziya. https://centrasia.org/newsA.php?st=1315851420. Date of the address: 01.07.2019
54. ria.ru [URL]. Terrorist attack at the Domodedovo airport on January 24, 2011. RIA Novosti. https://ria.ru/20140124/990669937.html. Date of the address: 02.07.2019
55. vedomosti.ru [URL]. Payment to the victims of terrorist attack in Domodedovo is one hundred times less than Bastrykin's assessment. Sheets. https://www.vedomosti.ru/business/articles/2016/06/07/644419-domodedovo. Date of the address: 05.07.2019
56. ria.ru [URL]. Terrorist attack at the Domodedovo airport on January 24, 2011. https://ria.ru/20160124/1363238677.html. Date of the address: 05.07.2019
57. ria.ru [URL]. Terrorist attacks on August 24, 2004 on the Tu-154 and Tu-134 planes. Reference. https://ria.ru/20090824/182146689.html. Date of the address: 05.07.2019
58. fb.ru [URL]. Plane crash in Egypt on October 31, 2015: reasons. Flight 9268. https://fb.ru/article/261934/aviakatastrofa-v-egipte-oktyabrya-goda-prichinyi-reys. Date of the address: 05.07.2019
59. gazeta.ru [URL]. "They seemed eternal": as in the USA the Twin Towers failed. http://gazeta.ru/social/2019/09/11/12641899.shtml. Date of the address: 05.07.2019
60. rusvesna.su [URL]. Destruction of "Boeing" of MH17: as it was. https://rusvesna.su/news/1583494279. Date of the address: 05.07.2019
61. gazeta.ru [URL]. Collapsed in air: as shahids destroyed Tu-134 and Tu-154. https://www.gazeta.ru/social/2019/08/24/12598531.shtml. Date of the address: 05.07.2019
62. ria.ru [URL]. Accidents of the Tu-154 plane and the Tu-134 plane in 2004. https://ria.ru/20140824/1021061881.html. Date of the address: 05.07.2019
63. birdinflight.com [URL]. 15 years later: The little-known facts about explosion of Tu-154 over the Black Sea. https://birdinflight.com/ru/mir/20161005-10-faktov-o-vzryve-tu154.html. Date of the address: 05.07.2019
64. og.ru [URL]. Explosion at the Domodedovo airport cost terrorists 500 thousand rubles. General Newspaper. https://og.ru/news/2012/01/24/58991. Date of the address: 05.07.2019
65. gov.spb.ru [URL]. Information on payments to victims in terrorist attack on April 3. Official site of the Administration of Saint Petersburg https://www.gov.spb.ru/gov/otrasl/trud/news/116979/. Date of the address: 05.07.2019
66. rbc.ru [URL]. Vagonmash in 2008 will deliver to the subway of St. Petersburg 27 cars for the sum about 491 million rubles. RosBusinessConsulting news agency. https://www.rbc.ru/spb_sz/freenews/5592c75b9a79473b7f4c32c4. Date of the address: 05.07.2019
67. fontanka.ru [URL]. What was shown by terrorist attack in the subway of St. Petersburg. Fontanka.RU. https://www.fontanka.ru/2018/04/03/008/?feed. Date of the address: 05.07.2019
68. kavkaz-uzel.eu [URL]. Act of terrorism at the Government house in Grozny on December 27, 2002. Internet mass-media "Caucasian knot". https://www.kavkaz-uzel.eu/articles/217876/. Date of the address: 05.07.2019
69. kommersant.ru [URL]. Terror in London. JSC Kommersant. https://www.kommersant.ru/doc/589452. Date of the address: 05.07.2019
70. iz.ru [URL]. Drones get off in packs. What problems were shown by attack of UAVs to Hmeymim base. News. https://iz.ru/693588/konstantin-bogdanov/drony-sbivaiutsia-v-stai. Date of the address: 08.09.2019
71. vestifinance.ru [URL]. 10 most expensive attacks of terrorists in the history. Online media "Messages. Economy (vestifinance.ru)". https://www.vestifinance.ru/articles/90806. Date of the address: 18.10.2019

Chapter 2
System of Anti-terrorist Safety of Airline

2.1 Concept of a Security System

The analysis and research of the reasons of realization of acts of terrorism in aircraft and consideration of main types of terrorist threats, allows drawing a conclusion that the modern and effective highly reliable system of anti-terrorist safety of airline has to resist similar aggressive manifestation. The system by the principle of functioning has to consist from some limited and, at the same time, enough the elements which are in communication with each other, at the same time, similar communication, maybe, as unilateral (in aspect of direct or indirect influence), and bilateral, to be carried out both as statement of conditions and requirements, and for interaction of elements for the solution of joint tasks. The existing normative legal acts did not establish the requirements to structure of elements of such system, as well as there is no concept of similar system that gives the chance to the author of the real research independently to develop the system covering aspects of ensuring protection against terrorist threats and to display it in the form of the conceptual scheme (Scheme 2.1).

Scheme 2.1 clearly demonstrates the structure of element base of system of anti-terrorist safety of airline. So, technical means of safety include the list of the used safety controls and protection to which carry: the television systems of observation, the security alarm system and the systems of the automated control and management of access, the automatic fire alarm system, the equipment of examination and control, various protective constructions and also warning systems. Subelements of human resources are faces on whose actions safety of airline directly or indirectly depends: service of aviation safety, staff of internal affairs, army of national guard, Federal Security Service, various private security companies and also the other persons who within the functions do not have direct tasks of safety (rather wide range of personnel—from technical specialists of airline to the commander of the aircraft). The element of organizational processes of airline is responsible for rather wide range of questions which it enters: internal standard regulation of activity, control and ensuring quality, management of activities for safety, selection and training, the choice of technical means of safety and also the general technological processes

P. P. Hachikyan, *Countering Air Terrorism*, Springer Aerospace Technology,
https://doi.org/10.1007/978-981-19-6172-4_2

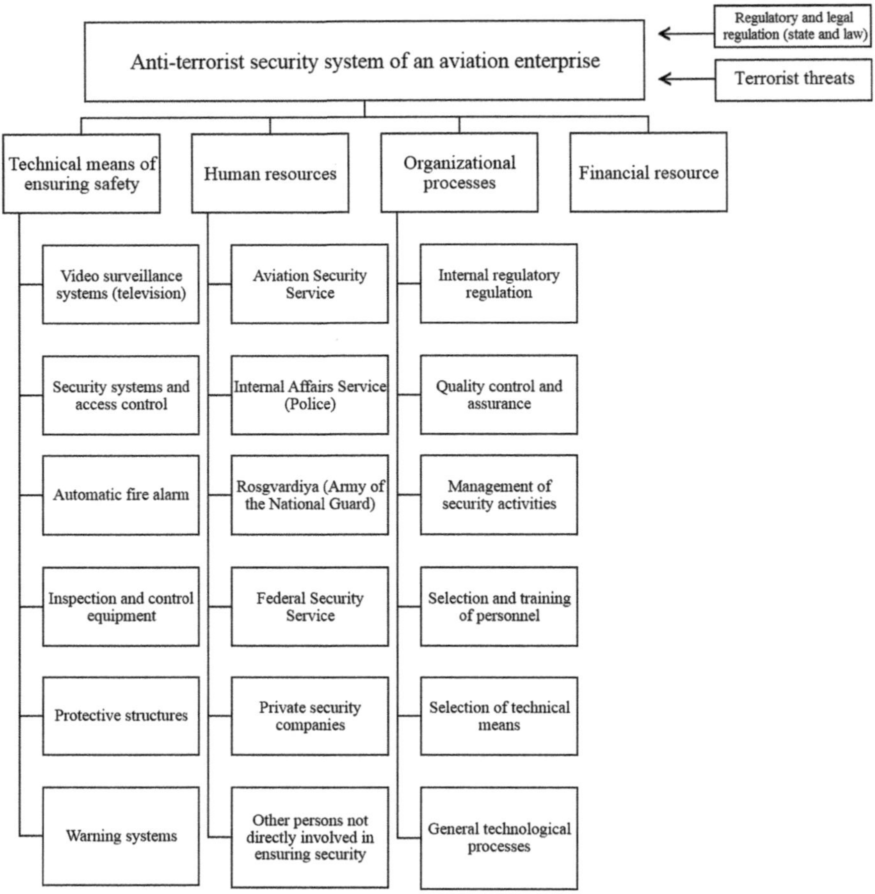

Scheme 2.1 The system of anti-terrorist safety of the air enterprise (author's conceptual scheme)

which are carried out at airline for questions of ensuring anti-terrorist safety (both direct, and indirect). The financial resource is allocated as a separate element, nevertheless—in the real research, it will be considered as the element accompanying practically each element and a subelement of system—so technical means, an indicator of their reliability of work without failures functioning—it is supposed to express as their cost, human resources is possible to estimate in finance—the salary, also, as well as organizational processes—in a projection of costs of their realization. Also two elements which though do not enter system are the major, but, render on it direct as favorable and fruitful—an element of standard and legal regulation (in aspect of the state, the law, the relevant international legal standards having both obligatory, and advisory nature), and destructive influence—terrorist threats in all types of their manifestations. At the same time, two given elements are extremely similar on the level of impact on system to those that categorically do not allow a message with them habitual interaction (as it happens to elements in the system) but only oblige to take influence for granted and to meet conditions (in relation to standard legal regulation) or to show to such element an extreme form of prevention, rejection and fight—in a case with terrorist threats.

The concept of carrying out scientific research is constructed on a research of the importance of the making *N-elements* of systems and development of solutions of decrease in risk of realization of terrorist threats in airline by improvement of the organization of productions and optimization of distribution of a financial resource at realization of system measures in relation to separate organizational measures of system of anti-terrorist safety of the air enterprise. Technical means of safety are considered in aspect *of the N-elements* providing reliability of functioning and representing the subelement system consisting of some number *of N-elements*. Human resources, also acting as highly reliable *N-elements* of system, considered from a position of their number, quality and reliability of work of experts, at the same time, the author develops methodical recommendations about improvement of quality of selection of employees about personal qualities and a selection method by probabilistic criterion. Corruption considered as the factor making an adverse and destructive effect on quality and reliability of work of experts. The financial resource acts as a measure of assessment both reliability, and quality of all *N-elements* of system for which distribution is developed the corresponding mathematical models allowing to make achievements of the set risk indicators at a certain budget. Models of system of ensuring anti-terrorist safety of airline and assessment of probability of act of terrorism developed for the objective proof of the importance of everything, without exception, N-elements of system and indicators of their reliability. The concept of an author's research is represented in Scheme 2.2.

The variety of elements and subelements of system given in Schemes 2.1 and 2.2 asks the corresponding circle of problematic issues, which is, in details, analyzed, investigated, and formed the basis for development and obtaining new scientific results that are shown in the subsequent chapters of the book.

2.2 Security System Elements

The need for officials of the air enterprises in due time to analyze and predict vulnerabilities for the purpose of prevention of possible incidents is extremely important for decrease in risk of realization of terrorist threat. Forecasting, as the prevention mechanism, is a paramount task at the organization of works in the field of safety as this activity allows choosing by means of mathematical modeling the most effective decisions. Proceeding from a problem of forecasting, for ensuring aviation and transport safety, the general anti-terrorist security, on objects of aviation infrastructure the measures directed to prevention of a bringing of the forbidden objects and substances are applied to defense of perimeter of an under control object from penetration of strangers. Providing measures carried out by application of organizational actions at which technical means of safety and human resources, which should be considered as N-elements (components) of system of anti-terrorist safety of airline, are involved.

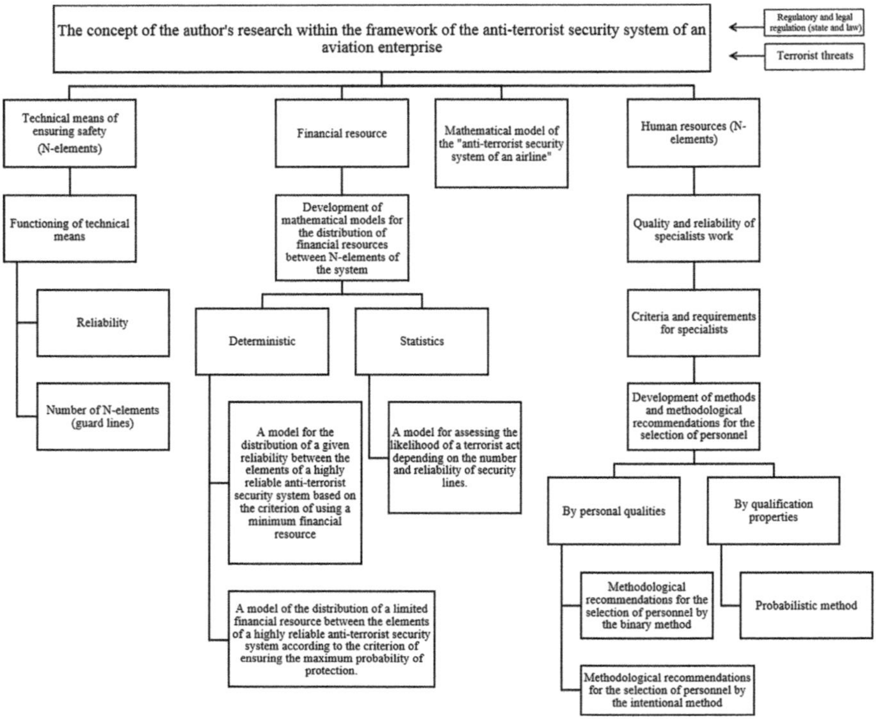

Scheme 2.2 Concept of an author's research

Considering the possibility of commission of terrorist attack, it is necessary to enter a concept risk *(Latin of resecu—"to cut", "reduce" or other—Greek "danger")*. The concept of risk strongly entered a lexicon of aviation safety that is confirmed by a position of the famous scientists-pilots: in the scientific works E. A. Kuklev applies risk—the focused approach to management of safety of difficult systems in the field of aircraft and also raises safety issues of flights on the basis of indistinct estimates of risks [1, 2], V. S. Shapkin, Dyomin S. S., etc. emphasize in the work the importance of risks in safety of flights and [3], B. V. Zubkov offers new original approach to assessment of risks, and V. D. Sharov emphasizes the importance of a concept of risk of the book "The Theory and Practice of Determination of Risks in Airlines at Development of the System of Management of Safety of Flights" [4], Y. M. Chinyuchin and E. N. Gipich raise the questions of "acceptable risk level" and methodology of management of risk on transport [5]. Questions of risk when ensuring aviation safety are also in detail stated in the profile manual "Aviation Safety" under B. V. Zubkov, S. E. Prozorov, S. I. Krasnov, V. M. Ilyin's authorship [6]. L. N. Elisov in the scientific works in detail takes up questions of mathematical modeling of aviation safety by means of a qualimetry where the questions of risk and vulnerability are also raised [7, 8]. It is necessary to emphasize that a concept of the risk within this research is considered as a combination of a possibility of approach of act of terrorism and the related adverse effects (the human victims, presence of victims, damage to the state and airline and so forth).

In a case with risks, the system of anti-terrorist safety of airline has to be in the level of acceptable (admissible) risk in which level is reasonable proceeding from safety requirements established by the top step of hierarchy for functional mission—ensuring transportation of passengers and freights by means of the air transportation, at the same time for holding customs procedures has to be minimized, providing necessary comfort for users of services—passengers. The acceptable risk always represents a compromise between the level of safety and opportunities for its achievement. The risk directly depends on the requirements of the top level of hierarchy of system (standard and legal regulation—the state and the law) which set its limit of admissibility. The author believes that within the real research, it is expedient to accept as acceptable risk of commission of act of terrorism, by analogy with accepted by ICAO for safety of flights, value $1 * 10^{-6}$, that is one act of terrorism can fall on one million flights. In spite of the fact that terrorist attacks are not always conducted directly on aircrafts and can be committed also in the territory of the airports and other objects of transport infrastructure of airlines, it is logical to assume that the binding of acceptable risk to the number of flights (a comment—on one million flights) is objectively true, this indicator is inseparably linked with a passenger traffic, quantitative value of number of people which uses services of aircraft and, in the course of flight implementation—inevitably passes inspection zones of airline, airport.

The increase in level of anti-terrorist safety that is not calculated properly can entail negative consequences in which operation of air transport for final recipients of services (passengers, clients carriers of commercial freight) will become customs procedures, unfairly expensive or absolutely uncomfortable in connection with difficulties of passing, and for carriers—economically unprofitable that, in turn, will lower appeal of aircraft as automobiles.

Considering questions of terrorist threats and types of terrorist attacks, costs of their commission, and costs of mitigation of consequences, it is necessary to draw an unambiguous conclusion that the main authorized objective of the terrorist organization is drawing by the action (terrorist attack) of the maximum damage at the minimum expenses. Proceeding from financing, material technical resources and presence of qualified personnel, malefactors choose various ways of commission of acts of terrorism. Ways and types of terrorist attacks directly depend on subject to commission of terrorist attack, level of its technical equipment and potential counteraction of security services, ultimate goals.

The research as highly reliable elements of system of ensuring anti-terrorist safety is supposed to consider the technical means involved in safety in the present. It is possible to assume that failure though one of elements (as well as its disoperation, functioning failure and so forth) will lead to the most destructive consequences—to the conducted terrorist attack. Rather general importance of a concept of reliability of the equipment involved in aircraft area needs to note that reliability is considered by scientists in the field of aircraft as one of key factors of decrease in risk of emergence of emergency situations [9, 10].

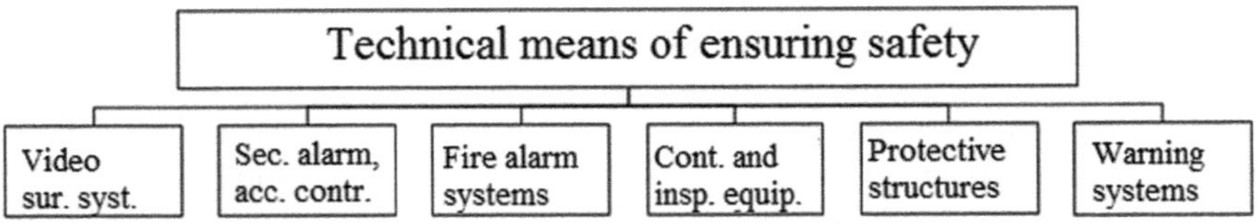

Scheme 2.3 Technical means—as elements of system of anti-terrorist safety of airline

Technical means—as elements of system of ensuring anti-terrorist safety, it is possible to structure in the following look (Scheme 2.3):

Further, for making decision on a possibility of reference of elements to a concept highly reliable, determination of their importance, briefly, we will consider technical means of safety:

- The system of video surveillance allows exercising remote visual control and protection of potentially vulnerable zones, places of a congestion of people, to carry out a photo and video fixing of the image with maintaining long-term archive of storage of information. Modern innovations in the field of development of similar systems allow to introduce algorithms of recognition of potential threats, to reveal suspects and to make automatic comparison of the image of the person with the identikit which is available in the database, to separate a huge array of static and unclaimed data and to provide to the operator system only information—a video series that is important for safety. The system of the device of lighting as the level of lighting allows increasing the clearness of recognition of details of the video image without use of expensive systems of night vision and infrared illumination having limited range also belongs to this section. The systems of video surveillance were widely adopted, are made by a set of firms manufacturers, characterized by the reliability at timely maintenance, are capable to carry out functioning without failures and malfunctions for a long time [11].
- Alarm systems, control and management of access are a complex of the equipment in which problem is remote control of a state and protection of potentially vulnerable zones of protection at which information on violation of zones (penetration on a certain territory, to the room) is visually displayed on the control panel and control of the operator in a convenient and clear look that gives the chance by means of the operators to exercise control of safety considerable on the area of the object. Development of technical means of protection allows to exercise control as behind certain rooms in the object, and to exercise perimeter control of territories on the open area.
- The equipment of examination and control included by a set of technical means for holding customs and control procedures for the purpose of detection of the forbidden objects and substances at pass of passengers and control of baggage and freights. Among the used equipment, it is possible to allocate metal detectors, radiological funds of examination, microwave scanners based on scanners of submillimetric waves and the system of remote detection of the forbidden substances based on the spectral analysis [12]. The special attention is deserved

by the technical means allowing to provide isolation and neutralization of the explosive device found onboard the aircraft. Similar devices are intended for the guaranteed protection of the aircraft against consequences explosion, ignitions, products of burning and other destructive factors under the influence of which the aircraft will be inevitable is destroyed or will suffer to a state at which will not be able to continue flight. Various explosion-proof containers and also devices of neutralization of electric detonators become really relevant in connection with achievements in a scope of various latest fire-resistant, anti-explosive, and neutralized materials now that is rather in detail described in works of scientists of B. V. Zubkov, S. E. Prozorova, etc. [13].

- The systems of fire safety include all complex of the funds allocated for early detection, suppression, and prevention of development of ignition on an object of aviation infrastructure. Are included in a complex of means: automatic fire alarm systems, warning systems and management of evacuation of people at the fire, the systems of automatic fire extinguishing, means of fire extinguishing (fireplugs, hydrants), a complex of technical actions (evacuation ladders, design decisions for creation of conditions for safe evacuation of people and prevention of development of ignition) [14].
- Protective constructions represent a complex of stationary and non-stationary objects, mechanical and electromechanical security systems of an object of aviation infrastructure, since its perimeter—fences (mesh, concrete, and other types), observant watchtowers and posts of protection, ditches and other blocking designs and also protecting barriers (gate, barriers, and barriers) to blocking of journey of motor transport. The researchers developed by the author protective constructions in the field of safety allowed to increase anti-terrorist security and reliability of security systems of the major state and international objects that also emphasizes a personal contribution of the author to a subject of the real research [15, 16]. The listed protective equipment are reliable designs and allow creating first and the boundary of safety, most obvious to the possible violator [17]. Such designs as a barbed wire, automatic locks of journey of transport the systems of counteraction to penetration can also call active.
- Warning systems represent the multi-purpose complex of the equipment directed to the solution of tasks of the sound and visual notification of the persons, which are in the territory of airline about emergence:

  (a) Emergency situations on the city line of the Ministry of Emergency Situations (Civil defense and emergency).
  (b) Ignitions in the building within functioning of means of the fire alarm system of an object.
  (c) Emergency situations, including at emergence and commission of acts of terrorism.
  (d) Need for broadcasting and submission of announcements of information and organizational character (about change of work of airline, about the lost children in the territory and so forth).

Warning systems can be as are integrated in uniform, and to consist from separate, depending on functional purpose and requirements imposed to them legislatively. Nevertheless, warning systems perform important function in the direction of actions of people in emergency and emergency situations and have to provide the help at evacuation from an object [14].

As well as it is shown on the Scheme 2.1, all above-mentioned technical elements log in anti-terrorist safety of airline. At assessment of their efficiency (as well as reliability) needs to believe about existence of linear dependence—than more financial resource invests in improvement of systems, especially their efficiency and ability to carry out objectives is high. The attention as well deserved by the fact that development of scientific and technical progress provides high growth of qualitative characteristics of technical means that positively affect the range of technical means of safety and their broad availability, to depreciation [18].

The analysis of technical means of system of ensuring anti-terrorist safety and personnel of airline objectively allows carrying these elements to the highly reliable *N-element* systems of anti-terrorist safety of airline, at the same time failure only of one of them (!)—can lead to terrorist attack commission.

Considering a question of increase in efficiency of anti-terrorist security of airlines, it is necessary to remember that terrorist attacks are conducted on air transport proceeding from the aspiration of terrorists to cause the maximum damage, to claim the greatest number of the human lives and also to have the maximum frightening effect on the world community the destructive actions. As a rule, at the organization of terrorist attack, the basic for terrorists is *the probability of successful commission of terrorist attack at the minimum investments, with use of shortcomings of safety of concrete airline.* The question of ensuring anti-terrorist safety of airlines at the same time concerns operators of aviation infrastructure around the world that generates an obvious question: "Why one states suffer from attacks of terrorists in activity in the field of aircraft more, than others?". Confirmed with statistics, the answer proved that *the least protected* airlines whereas aspect of the state of location of airline, nationalities of the potential victims, etc. remains on the second plan *are subject to the greatest threats.*

Considering questions of ensuring anti-terrorist safety in the USA, it is necessary to return once again to notorious events on September 11, 2001. Defects of holding customs procedures allowed an opportunity to persons with illegal intentions to rise by boards of aircrafts, and lack of harmonious and operational work of security services allowed occupying control of aircrafts in air that is gross violation of rules of safety. The whole series of mistakes, which is confirmed by the number of at the same time seized aircrafts, and almost identical algorithm on which terrorists carried out the artful plan led to the emergence of a tragic situation, i.e. the mistake had character of a system defect. Considered prior to this incident, rather reliable system of safety in aircraft of the USA showed the inability to resist to threats. The specified events transferred work of security services to a condition of the increased readiness that entailed enormous expenses, but, nevertheless and it appeared a little: it is enough to give the only example showing low efficiency of the taken measures—in September, 2003, the American student for an experiment could bring on a board of aircrafts at two airports (including the capital airport Baltimore) things, similar to those which were used by terrorists at events 11-09: matches, knives for cutting of a cardboard, bleach in tubes from under foundation [19]. It should be noted that still until recently, in the USA rather simple method of work of security services at which security services equally treated examination of passengers, without special attention to the identity of the passenger and ethnic origin for respect for tolerance applied. Over time algorithms of work of system of the preliminary analysis of information about passengers that contains information on need to pay special attention to these or those persons, but, nevertheless, accurate installations on carrying out psychoanalytic work directly with the passenger as with the person—services of aviation safety in the USA have no were developed [20].

The system of ensuring anti-terrorist safety in Great Britain is also not deprived of shortcomings—in 2006 to plans of the Al-Qaeda terrorist organization * exclusively preliminary work of police and special services whereas security measures were in the airports at the usual level, prevented to carry out capture and explosion of 20 aircrafts [21]. The prevented series of terrorist attacks caused, eventually, increase in level of safety at the airports of Great Britain to critical, at the same time on points of examination, considerable turns of passengers were formed, as temporary measures transportation of hand luggage was forbidden and also the list of the objects allowed for baggage is considerably changed. Despite a so-called "Muslim trace" concerning most of the persons who are earlier detained on suspicion of preparation of terrorist attacks, special algorithms on control of similar ethnos it was accepted not, in connection with continuous stay, considerable on number, and stay on lawful conditions in Great Britain of foreigners and immigrants from the countries of the East.

Ensuring anti-terrorist safety in aircraft of the People's Republic of China (PRC) is of keen interest. Considering considerable population, the People's Republic of China is possible to hold a position of one of the safe countries in the field of ensuring transport safety. As the bases of work of system of safety, it is necessary to allocate: practice of escort of aircrafts by law enforcement officers, a high competition on replacement of vacancies in law enforcement agencies (to three thousand people on one place), the special mechanism of interaction of law enforcement agencies with civil society regarding anti-corruption which provided a possibility of the state of material encouragement of the citizen who reported about corruption [22]. It is also worth selecting and administrative decisions: law enforcement agencies of the People's Republic of China have double subordination—to the Ministry of public safety of the People's Republic of China through operational and service activity, to the Ministry of Transport of the People's Republic of China through material support and on organizational and regular questions, for maintenance of mobilization and operational readiness of divisions of police proceeding from real requirements of branch [22]. Considering the stated data, it is admissible to draw a fair conclusion—the state is interested in involvement in safety of all population of the country by means of increase in prestige of law enforcement agencies and effective measures for the stimulation of citizens in questions of anti-corruption. The measures of technical character applied in the People's Republic of China differ to the front lines, innovative character: for counteraction to terrorism and identification of suspects the systems of recognition of suspicious criminogenic elements on emotions are introduced, and identification of the persons brought in a card file becomes possible to carry out on an arrangement of large veins and capillaries on a hand to within recognition of one hand from one billion at the processing time spent for it system no more than the three tenth fractions of a second [23].

The special attention is deserved by ensuring anti-terrorist safety in Israel. According to statistical data, the Ben Gurion airport located in the capital of Israel—Tel Aviv and national airline—Al's Ale airline is the safest in the world [24]. The first place in the field of ensuring aviation safety Israel deserved thanks to the system of work directly with passengers whereas it any innovative solutions in the field of the technical and customs equipment is not applied. Studying of the alleged passenger was carried out even long before his arrival in the airport, so at the time of purchase of tickets, check on existence of negative information on the passenger in databases of law enforcement agencies made. Detection of information of similar character will inevitably lead to more careful examination of the specific passenger and the persons, which are together with it. At an entrance to the airport special checkpoints at which external survey of the car and the being persons can be also made are equipped, the general visual control, including for psych emotional condition of the persons which are inside is made. Staff of security services is as on customs points, and masks in civil clothes among other passengers, for carrying out the covert surveillance and identification of suspicious subjects. There are data on regular support of security services of boards of aircrafts by employees and also on features of designs of vessels of the Israeli airline Al's Ale regarding compliance to all possible requirements of ICAO for safety of a cabin of pilots. At the heart of effective (but, in too time still not really unusual for Russian Federation, for example) works of security services with passengers use of so-called profiling (from English word "profile") is put [25]. The essence of a similar way consists in carrying out operational visual and verbal assessment by the security service specialist—references of the passenger to this or that profile (type) for assessment of its reliability, a possibility of commission of illegal acts. The qualified expert-profiler, in the theory, can try to find out true intentions of the passenger in several minutes and to distinguish a lie and doubts, at the same time a basis for the analysis will be only the appearance, a mimicry, gestures, speed of issue of answers to questions of the passenger and others, at first sight insignificant, habits and features of the specific person [26]. Considering extremely compressed time for carrying out the similar analysis of the passenger, the main objective for the security service specialist is to distinguish an obvious lie and confusion in the passenger, at the same time, his appearance and ethnic origin will also be an element at creation of the general opinion in the opinion of the security service specialist if she is previously correlated to the terrorist's profile (for example—wearing traditional national clothes, headdresses and so forth). Respectively, in that case the staff of security services has to be not just qualified security guards, and to have the skills of psychological work and special modern methods of psychoanalysis allowing to identify precisely and quickly threat hidden in the lump of people. In case of existence of certain suspicions, the suspicious passenger and his satellites exposed to careful examination in a sterile zone of the airport. Nevertheless, if passengers use services of air transportation rather often, they can choose for themselves to use of so-called cards of the checked client which automatically categorize the persons as the persons posing a small threat who received them. Obtaining similar maps is possible only to citizens of Israel who beforehand grant permission for verification of the biographic data in a voluntary order and lead a discussion with representatives of security services of the airport. Undoubtedly, the considerable staff of security services, emergency measures applied to safety in aircraft of Israel demand also the corresponding financial expenses, so for maintenance of the general safety of the airport the considerable sum of 25% of the budget of Tel Aviv annually allocated [24].

Proceeding from international experience, Israel is allocated with the most attentive relation to a question of the personality and the person, which is shown as selection of personnel, and use of modern psychological receptions during the work with passengers. It should be noted that the procedure of selection of shots for services of aviation safety is similar to the procedure of a profiling of passengers—a question only that on the procedure of selection it is possible to find considerable time whereas at a profiling—time is limited several minutes, and importance of the similar decision is extreme is important. Similar approach to bigger attention of the personality and the person well proved that confirmed by statistics of aviation incidents, which demonstrate lack of any considerable aviation incidents of terrorist sense in Israel for the last decades. In comparison with Israel, the demand of techniques of a profiling in Russia appeared rather recently, and continues to grow. It is also possible to note and obvious tendencies of demand of similar techniques, which allow to turn them from the means used only by people—into the automated systems of artificial intelligence,—so since February 1, 2020 in Russia GOST P 58776–2019 "Monitors of behavior and forecasting of intentions of people for the first time took effect. Terms and definitions", developed by Peter the Great St. Petersburg Polytechnic University, which, including, defines possibilities of work of the systems revealing the persons menacing to safety based on algorithms of artificial intelligence [27]. People's Republic of China allocated as the most densely populated country with high growth rates of technical means. To report the best practices of anti-corruption based on interest of the respectable citizen about the corrupt official, supplements a positive picture of the country in which questions of providing an order, safety and legality come to the forefront. In the People's Republic of China also, as well as in Israel special attention is paid to a question of the person, personality as to fundamental elements of the creation of a reliable modern security system.

2.3 Conclusion of the Chapter

The above also allows drawing rather valid conclusion that ensuring anti-terrorist safety of airline leans on three nodal factors to which it is necessary to carry: technical means of safety, staffing of security services and the general organization of work at airline, a financial resource.

As for overall performance of technical means of safety, here to the forefront, there is the cost of these means—their financial equivalent. The assumption of almost linear dependence of cost of technical means (in decibel scale) and efficiency of their work and reliability as which quantitative measure *the probability of detection* of the carried by undesirable object or substance can act looks rather natural. Such situation objectively generates a task about distribution of a limited financial resource between different types of equipment.

Accurate performance of the functions and instructions aimed at providing anti-terrorist safety from the staff of airline and law enforcement agencies, most, depends also on their material security. However, in this case, it is essentially about especially nonlinear dependence. This fact of nonlinear dependence generates a tendency of rapid reduction of reliability of the worker from the appearing sums of money (a bribe, a corruption component). The certified personnel—the military personnel (including the staff of FSB) and the staff of Department of Internal Affairs (police, internal service) and also Rosgvardiya's employees are involved in ensuring anti-terrorist safety of airline as civil personnel (SAB, PSF, the workers of objects of aviation infrastructure who are not connected with safety of flights) and also. The person, also, as well as technical means, objectively acts as an element of systems of anti-terrorist safety of airline. Studying of international experience fight against terrorism on transport showed that the most reliable systems of counteraction to terrorism from the point of view of safety are applied in Israel and DPRK, at the same time, despite improvement of material and technical resources of security systems and customs complexes, the specified states into the forefront ask personnel selection of persons responsible for safety and also applicable them techniques of identification of suspicious subjects. It puts the selection of the persons involved in ensuring anti-terrorist activity, their personal qualities, internal aspirations, and the moral principles that are, in detail, described in the present book further in the forefront.

The above allows drawing a conclusion on the need for the optimization of work of a number of elements on which the system of ensuring anti-terrorist safety of airline based:

(a) The human resource involved in ensuring anti-terrorist safety of airline.
(b) Technical means of ensuring anti-terrorist safety of airline.
(c) Various organizational processes concerning as directly points (a) and (b), and the general organization of the activity of airline exerting impact on questions of ensuring anti-terrorist safety.

(d) Financial resource.
As the elements making direct impact on system, it is necessary to allocate:
(e) Standard and legal base (regulation from the state, the top level of hierarchy).
(f) Terrorist threats.

References

1. Kuklev EA (2019) The risk-focused approach to management of safety of difficult systems on the basis of fuzzy logic of the analysis of criticality of operational situations. In: XIII All-Russian meeting on problems of management VSPU-2019, Moscow on June 17–20, 2019
2. Kuklev EA (2016) Management of safety of flights of aircrafts on the basis of indistinct estimates of risks of emergence of emergency conditions of flights. Scientific Bulletin of the Moscow State Technical University of Civil Aviation, no 226, pp 199–205
3. Shapkin VS (2017) To a question of application of risk-oriented approach in a problem of safety of flights. In: Shapkin VS, Dyomin SS, Nikitin AV, Dyomin DS, Kovtushenko DVM (eds). Scientific Bulletin of State Civil Aviation Research Institute, vol 16, no 327, pp 61–71
4. Zubkov BV (2010) The theory and practice of determination of risks in airlines at development of the system of management of safety of flights. In: Zubkov BV (ed) Spheres of EL of M. MSTU of HECTARE, 196 pp
5. Gipich GN (2010) Introduction to the theory of risks. In: Gipich GN, Chinyuchin YuM (eds). Scientific Bulletin of the Moscow State Technical University of Civil Aviation, no 160, pp 7–11
6. Zubkov BV (2014) Aviation safety. The textbook for students of higher educational institutions (under the editorship of Prozorov SE). In: Zubkov BV, Prozorov SE, Krasnov SI, Ilyin VM (eds). FGBOU VPO "The Ulyanovsk Highest Aviation School of Civil Aviation (in-t)", Ulyanovsk, 410 pp
7. Elisov LN (2006) Kachestvo of vocational training of aviation personnel and safety of air transport. Monograph. ITsPKPS Publishing House, 244 pp
8. Elisov LN (2017) Aviation safety as subject to mathematical modeling. Scientific Bulletin of the Moscow State Technical University of Civil Aviation, vol 20, no 3, pp 13–20
9. Vorobyov VG (1995) Reliability and efficiency of the aviation equipment. In: Sparrows VG, Konstantinov VD (eds) The textbook for higher education institutions. Transport, 248 pp
10. Vorobyov VV (2017) Role of the radio-electronic equipment in management of safety of flights at implementation of the Global air navigation plan of ICAO. V.V., A.V. Vlasov's sparrows. Scientific Bulletin of the Moscow State Technical University of Civil Aviation, vol 20, no 4, pp 156–161
11. Hachikyan PP (2020) About the prospects of development of systems of video surveillance at implementation of measures for ensuring anti-terrorist safety of airline. Sci Meridian Online Mag 1(35):261–263
12. Hachikyan PP (2017) About some ways of preflight radio-electronic examination of baggage and passengers. Qual Life 2(14):111–116
13. Zubkov BV (2008) Development of methods and means of protection of the aircraft from threat of explosion during flight. In: Zubkov BV, Prozorov SE, Bochkaryov AN, Zenkov ASM (eds). Scientific Bulletin of the Moscow State Technical University of Civil Aviation, no 127, pp 139–145
14. Hachikyan PP (2020) Sistema of fire safety and the emergency notification—as the most important element of creation of a complex of anti-terrorist protection of airline. Sci Meridian Online Mag 2(36):462–464

15. Hachikyan PP. Guard cabin: patent for the invention of RU2586992C1 Russian Federation: MPK E04H/Hachikyan of Items—No. 2015109246/03; 17.03.2015; published. 10.06.2016, bulletin No. 16
16. Hachikyan PP. Observation and watchtower: patent for the useful RU163226U1 model Russian Federation: MPK E04H/Hachikyan of Items—No. 2015109241/03; 17.03.2015; published. 10.07.2017, bulletin No. 19
17. Hachikyan PP (2020) About some questions of application of protective constructions at creation of systems of anti-terrorist safety of airline. E-Scio 1(40):116–124
18. Hachikyan PP (2019) Philosophical aspects of a role of scientific and technical progress. In: The collection: innovative technologies in science and education the collection of articles XIV of the international scientific and practical conference, Penza, 2019, pp 97–100
19. Belov PG (2006) Methodological bases of ensuring transport safety. Transp Saf Technol 3(8)
20. Demidov YuN (2006) Features of visual psychodiagnostics of the identity of the terrorist. Detection of criminal distortion of information. In: Demidov YuN, Anisimova NN (eds) Studies—a method. Grant. VIPK Ministry of Internal Affairs of the Russian Federation, Domodedovo
21. iz.ru [URL]. In London terrorist attack, similar is prevented "on September 11". 20 terrorists detained. https://iz.ru/news/383501. Accessed 02 Nov 2019
22. Alexandrov AN (2017) Features of safety on transport objects in People's Republic of China. In: Alexandrov AN, Petrykin NN (eds). Bulletin of the East Siberian Institute of the Ministry of Internal Affairs of the Russian Federation, vol 4, no 83, pp 16–22
23. tass.ru [URL]. The airports of China began to use technology of recognition of emotions and capillaries on a hand to within recognition of one hand from one billion at the processing time spent for it system no more than the three tenth a fraction of a second. https://tass.ru/obschestvo/7069902. Accessed 02 Nov 2019
24. Zhukov SI (2006) Use of the international experience in safety and law and order in the territory of the airport (on the example of the Ministry of Internal Affairs of Italy). Yurist-pravoved, no. 5
25. Kudin VA (2013) Profayling in activity of Internal Affairs Service: from the theory and methodology to practice. In: Kudin VA, Stately VM (eds). Bulletin of the St. Petersburg University Ministry of Internal Affairs of the Russian Federation, vol 3, no 59, pp 4–15
26. Karagodin AV (2014) To a question of efficiency of a profayling at implementation of customs actions on objects of transport infrastructure. In: Karagodin AV, Stolbina LV (eds) Problems of law-enforcement activity, no 3, pp 24–27
27. gost.ru [URL]. ROSSTANDART. Federal Agency for Technical Regulation and Metrology. The first standards in the field of artificial intelligence are approved. https://www.gost.ru//newsRST/redirect/news/1/6566. Accessed 28 Dec 2019

Chapter 3
Mathematical Models of Increase in System Effectiveness of Anti-terrorist Safety of Airline

3.1 Mathematical Modeling of a Security System

For achievement of the object, set in a research, the solution of a scientific task of distribution of a financial resource between elements of the system of anti-terrorist safety of airline is required. As it is shown in the book earlier, the system of anti-terrorist safety of the air enterprise includes a number of elements and sub-elements to which certain functional tasks are assigned. By the researches conducted in the first and Chap. 2, it is objectively established that technical means of safety, a human, and financial resources are among the elements of a highly reliable N-element system of anti-terrorist safety of airline. At the same time, the element of organizational processes is rather generalized, and can be conditionally divided into sub-elements which correspond to human resources and technical means, respectively, that do not demand separate consideration as the element which is separately allocated earlier in work. The need of distribution of a financial resource is caused by the existence of a positive element of influence—the set requirements of the top level of the hierarchy of system—the standard-right regulation for branch stated by the state, the international normative legal acts, and also the existence of a negative element of influence—terrorist threats, at all this, there is a limited financial resource (the budget of airline) which does not allow to fulfill each requirement (in a projection to characteristics of elements) in the best way (providing the set safety level).

For the solution of this scientific task, the most rational is the application of a mathematical apparatus to the modeling of distribution of a resource by various criteria that can be represented conceptually in the following look (Scheme 3.1).

P. P. Hachikyan, *Countering Air Terrorism*, Springer Aerospace Technology,
https://doi.org/10.1007/978-981-19-6172-4_3

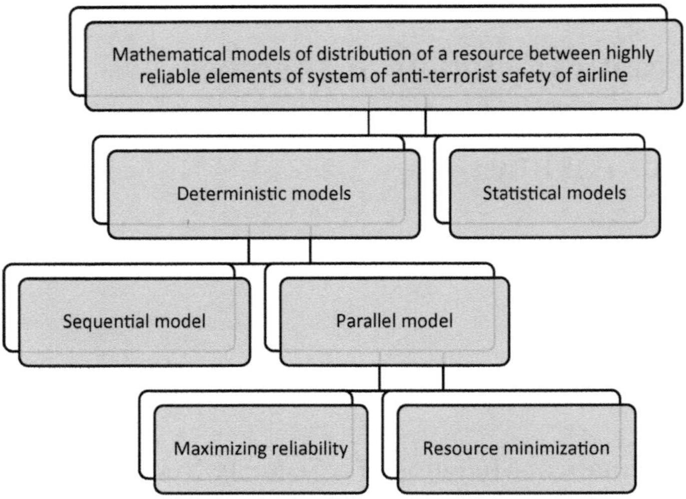

Scheme 3.1 Mathematical models of distribution of a resource between highly reliable elements of system of anti-terrorist safety of airline

3.2 Deterministic Models of Distribution of a Financial Resource Between Highly Reliable Elements of System

Application of deterministic models in the real research is most justified in that case when parameters and variables are certain (nonrandom) sizes. For the creation of models, knowledge of behavior and interaction of all elements of the modeled system are used. Similar models are attractive to the analysis thanks to presentation and usability, give an opportunity to obtain necessary and sufficient information for decision-making that is especially relevant in this research.

3.2.1 Consecutive Model

As it was established above, the system of anti-terrorist safety of airline, it is possible to present in the form of uniform structure *of S* consisting generally of some, rather large *number N* of elements, each n_i, where $j = 1, 2\ldots N$, from which, they can be independent of each other (Scheme 3.2).

As *N-elements*, it is necessary to consider, but not to be limited, the following:

1. $_{Nsld}$—the element which is responsible for the implementation of entrance examination of staff of airline and law enforcement agencies on the air terminal territory (a security zone of the first type).

Scheme 3.2 Structure of S with a set of N_j elements

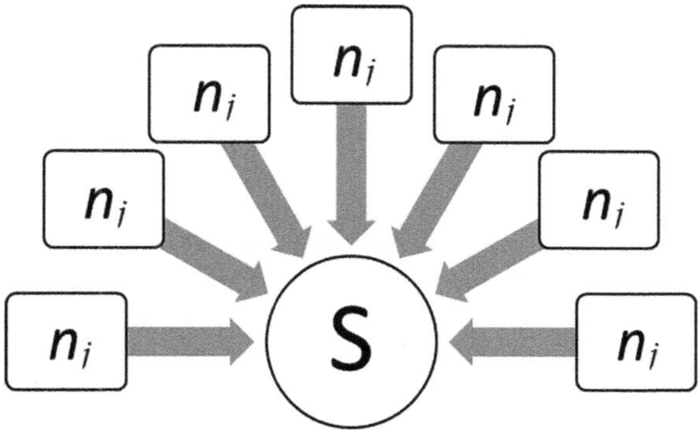

In the course of carrying out entrance examination of passengers and baggage, technical means of safety (customs complexes) and also directly the personnel of service of aviation safety of the enterprise exercising control are involved.

2. N_{pd}—the element which is responsible for carrying out a preflight examination of passengers and baggage just before landing aboard the aircraft (a security zone of the second type).

   At a certain identity about N_{vd}, *the Npd element*, differs in the existence of more innovative equipment of safety (customs complexes) both a large number and quality of the persons involved for the organization of this action from outside of airline according to the existing normative legal acts.

3. N_{vd}—the element which is responsible for the implementation of entrance examination of passengers and baggage at the primary pass on the air terminal territory (a security zone of the third type).

   In the course of carrying out entrance examination of passengers and baggage, technical means of safety (customs complexes) and also directly the personnel of service of aviation safety of the enterprise exercising control are involved.

4. N_{komgr}—the element which is responsible for carrying out an examination of exclusively commercial freights, mails, and onboard stocks of the aircraft.

   Undoubtedly, it is impossible to exclude a possibility of commission of act of terrorism without the presence of the terrorist onboard air when the explosive device is among the moved freight. Due to the insufficient competence of personnel of airline, commercial freights, mails, and onboard stocks, in difference from hand luggage, can undergo not such careful control. It is also necessary to pay attention to a possibility of the biological threat of infection of onboard stocks (water and food) for poisoning of crew of the aircraft and passengers as extraordinary, but in too time possible act of aggression from terrorist communities.

5. N_{tsb}—the element which is responsible for complex providing airline with technical means of safety.

The following subsystems enter a complex technical means of the safety of the air enterprise:

- television systems of observation;
- the alarm system and the automated control and management of access;
- systems of fire safety and notification;
- equipment of examination and control;
- protective protections.

The listed systems make the centralized system engineering decision realizing the main objectives in the field of prevention of approach of emergency and extraordinary events—threats to security, and in case of their approach—provides the notification and evacuation of people, promotes investigation of the occurred incidents (by means of existence of an archive of events, video images, records, and so forth).

6. $_{Nobsluzh}$—the element which is responsible for safety of consequences of holding procedures of repair and service of aircrafts and air terminals at which not operability to the terrorist it becomes possible:

   - to make intended damage (substitution on obviously faulty) the equipment (flight, navigation, and so forth) the aircraft (air terminal) which will entail emergence of a situation, the crash of the aircraft, and other irreversible consequences;
   - to hide in the cunning way in the aircraft (on air terminal) the explosive device, control of which can be exercised both from the timer and at receipt of a regular signal of the aircraft (at the achievement of a certain height, at the change of a trajectory of the movement, and so forth);
   - by means of the regular equipment of airline to make accident on a runway with the aircraft;

7. $_{Nchelres}$—the element which is responsible for human resources—the personnel who are taking part in ensuring anti-terrorist safety.

   The possibility of the terrorist to use the existing shortcomings of a security system or to create new depends on the personnel and its qualities.

   It is necessary to refer the following categories of employees to personnel:

   - employees of SAB of airline;
   - the staff of the Department of Internal Affairs on transport (Ministry of Internal Affairs);
   - Rosgvardiya's employees;
   - staff of the border service (FSB);
   - personnel of private security companies.

8. $_{Nbpla}$—the element which is responsible for a possibility (technical equipment, existence of the specialized equipment, shots) of airline to resist to threats from the unmanned aerial vehicles operated by terrorists.

   The question of detection and counteraction to unmanned aerial vehicles is relevant now. The small aircraft operated remotely, and possessing an advance programmed route can bear explosive devices on the board, direct hindrances in a radio communication, and, in general, are dangerous weapons in hands of terrorist communities due to the lack of universal and effective remedies of counteractions.

9. N_{morust}—the element which is responsible directly for moral stability of the personnel who are taking part in ensuring anti-terrorist safety.

   Moral stability is the difficult social and psychological and moral characteristics of the employee which directly influence his abilities:

   - to resist to corruption temptation in the form of bribery from terrorists;
   - to resist to temptation to commission of illegal acts, assuming impunity (rejection by the employee of the unconditional principle of inevitability of punishment);
   - to make self-sacrifice the highest form of manifestation of altruism. Ability to put the interests of work (service) and society (life of other people)—above personal interests, even if the employee risks own life and/or life of the family.

10. N_{dvprov}—the element of "double check" which is responsible for a possibility of commission of an intended or inadvertent error of one element of system or the employee without its control (elimination) by the second element of system or the employee.

    It is necessary to consider a similar element as the most important instrument of control and check, which allows eliminating both intended and inadvertent (casual) errors of technical systems and personnel. At the same time, for theoretical increase in reliability, is important to consider functioning of any security systems at which they have to possess not less than two levels of check. As an indicative example of such level of double check, it is possible to call passing of baggage through X-ray the device of system of examination at which—in case of lack of the mechanism of double check, the forbidden freight can be passed at bribery by terrorists of only one employee making examination whereas at the mechanism of double check, freight on the screen of customs installation is watched by two security service specialists at once (at the same time, one of them can is at distance, in the special room and has no possibility of contact with colleagues and passengers) that, in general, it is designed to reduce chances of a bringing of the forbidden subject or substance, and, respectively, increases efforts (expense) of terrorists to carrying out terrorist attack, but—affects cost.

11. $N_{obkhoda}$—the element of "round" implying a possibility of terrorists to bypass the established safety rules of the air enterprise, by the organization of various non-standard (not widespread) methods of penetration and a bringing of the forbidden objects and substances on the protected object, such as:

    - use of counterfeit documents of expeditious cover and certificates of staff of federal executive authorities, the special services allowing to get into zones of examination and the territory of airlines without holding customs procedures;
    - involvement in criminal activity of staff of the MFA—the diplomatic personnel and diplomatic couriers transporting diplomatic mail and freights

which is not subject to opening and examination according to the international agreements;

- imitation and creation of emergency situations (a critical medical condition of the passenger, accident of technical systems, and so forth) in security zones of airlines for the purpose of penetration on an object of terrorists under the guise of service of ambulance, rescue service or repair crew which authenticity and true motives of actions will not be checked and be established by service of aviation safety at the time of their emergence in the territory of airline due to the need urgently to liquidate consequences;
- various schemes of introduction, the operated by terrorist communities, commercial organizations in daily activity of airline (the conclusion between the organization, affiliated with terrorists, and airlines of contracts—on service of engineering systems of airline, on cleaning of the territory, delivery of onboard food and so forth), for the purpose of providing easy access of terrorists in the protected zones of airline without suspicions from security services.

Let's assume that in case functioning of one of the above-stated elements depends on the work of another, we will consider them as one element. At the same time again we will highlight that failure only of one element of system can have the most catastrophic consequences in the form of the realized act of terrorism.

Let's make one more assumption that all elements entering into structure of S represent *highly reliable* systems. Highly reliable system it considered such system which reliability not less set and for which qualitative criteria of reliability are met.

It is important to note what by this consideration does not play any role that specifically is system elements—the equipment, people, qualification, the system of the organization and management, etc.

Probability that during T time *the element* n_i will function without failures we will designate as P_j, at the same time $j = \overline{1, N}$.

As for T time during which the element has to function normally, this time is set by "the top level of hierarchy" in relation to the checked elements.

It is expedient to treat and call expression of $R_j = (1 - P_j)$ risk. It is necessary to pay attention that the risk is function of time, but that is extremely important that during the required time the probability of P_j did not decrease.

At the accepted designations probability of successful work, all N elements of system S will be obvious to be defined by equality:

$$P_{\Sigma} = 1 - \prod_{j=1}^{N} (1 - P_j), \tag{3.1}$$

it is necessary to have to a look that values of each of P_j are close to 1.

The author believes that within the real research, it is expedient to accept as acceptable risk of commission of act of terrorism, by analogy with accepted by ICAO for safety of flights, value $1 * 10^{-6}$, that is 1 act of terrorism can fall on one million flights. In spite of the fact that terrorist attacks are not always conducted directly on aircrafts, and can be committed also in the territory of the airports and other objects of transport infrastructure of airlines, it is logical to assume that the binding of acceptable risk to the number of flights (a comment—on one million flights) is objectively true, this indicator is inseparably linked with a passenger traffic, that is quantitative value of number of people which uses services of aircraft and, in the course of flight implementation—inevitably passes inspection zones of airline, airport.

Further, as it accepted in ICAO, we will present probability P_j in the form:

$$P_j = 10^{-n_j}. \tag{3.2}$$

Then according to (3.1),

$$P_\Sigma = 1 - \prod_{j=1}^{N} \left(1 - 10^{-n_j}\right) = 10^{-M}. \tag{3.3}$$

Expression for risk that is for the probability that, at least one element, did not execute the function will have in this case an appearance:

$$R_\Sigma = 1 - 10^{-M}. \tag{3.4}$$

For the considered tasks, it is represented to assume naturally that P_j has values, at least, more than 0.97–0.99. Risks, in this case, do not surpass units of percent, therefore, parameters n_j have values about 0.005–0.013.

On Schedule 3.1, the dependence of size of risk of R_Σ of system on the number of identical highly reliable elements, probability of performance of the functions by them, respectively, is shown there was 0.97, 0.98, and 0.99.

Essential growth of risk at an increase in the number of elements attracts attention. So even for very highly reliable elements ($P = 0.97$) at number of elements 12 reliability of system falls by two times.

Schedule 3.2 represents similar dependence for a situation when from reliability positions all elements are not identical and differ from each other.

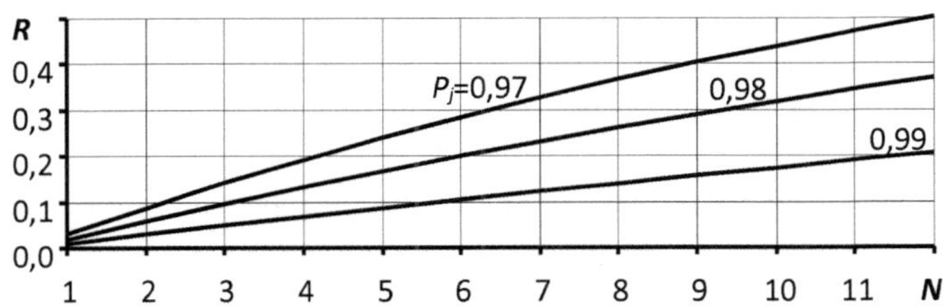

Schedule 3.1 The dependence of risk $\mathbf{R}_\Sigma$ on the number of control elements N with their different reliability probabilities $P_j = \text{const}$

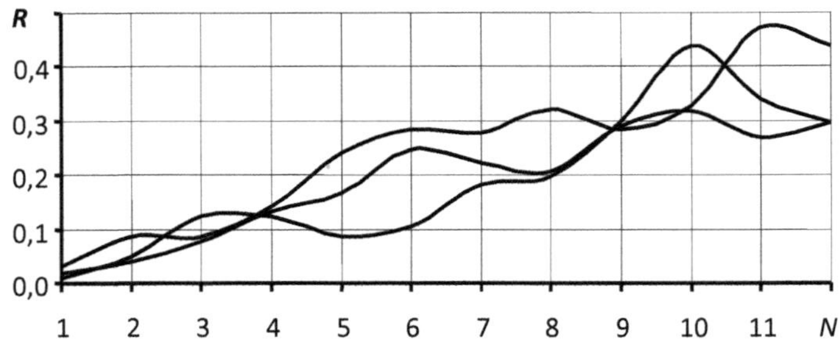

Schedule 3.2 The dependence of risk $\mathbf{R}_\Sigma$ on the number of control elements N with their different reliability probabilities $P_j = \text{const}$

In absolute majority of cases, an increase in the reliability of performance of the tasks by any element including the person who is a part of the system ensuring the protection of anti-terrorist safety of airline comes down to allocation of the appropriate financial means of *d*. Of course, this statement cannot be understood in the absolute sense of value. Nevertheless, such general tendency is looked through unambiguously. At the same time, financial means can be spent as it is one-stage (purchase of expensive customs complexes, reconstruction of objects, and so forth) and in the form of long-term monthly investments to which it is necessary to carry—the salary fund of employees of security services, costs of maintenance of security systems, buildings, constructions.

Of course, at such interpretation of function, the connecting risks with a financial resource—$f_j(d_j)$, it is hardly possible to treat as smooth and continuous, most likely, they should be considered as step, spasmodic. However, for the further analysis, which is not demanding the detailed quantitative analysis as the interest will be is of by qualitative tendencies of succession of events, use of smooth continuous functions is quite acceptable. Further the $f_j(d_j)$ functions, we will call *functions of reliability*.

Rather obvious, but very essential conclusion follows from the ratios received above. The risk is less, or the parameter is less n_j, the importance is more, it is possible to tell cost, *j*-oho an element at stages of its use and at a creation stage. At the same time, that extremely essentially, the dependence of cost of s_j of each element of system on the size of risk r_j has especially nonlinear character.

The representations received above open an opportunity to formulate a problem of distribution of a financial resource: "How in the system consisting of N mutually independent elements providing some initial risk of R_0 to distribute an additional financial resource of D between these elements to provide the minimum value of risk?".

Question, nodal at such problem definition, is development of model of cost, that is creation of function $d_j = f(n_j)$.

The feature of this task consists that function C has to be considered in the neighborhood of zero as highly reliable systems for which as it was shown above, n_j it is close to 0 are considered. At the same time, that rather obviously, in a point $n_j = 0$ required function has a pole of some order q^j.

Told allows to present function $f(n_j)$ in the form of Loranovsky decomposition [1]:

$$f(n_j) = \sum_{i=1}^{q_j} \frac{C_{-i}^j}{(n)} + C_0 + \sum_{i=1}^{\infty} C_i^j n_j^i, \tag{3.5}$$

where $C_{\pm i}^j$—some coefficients of decomposition.

As $\frac{C_{-(i+1)}}{n_j^{i+1}} \gg \frac{C_{-i}}{n_j^i}$, the main contribution to the sum (3.5) brings composed the first sum with the index m_j. In other words,

$$f_j(n_j) = \frac{C_{-m_j}^j}{(n_j)}\overline{m_j}. \tag{3.6}$$

Equality (3.5), naturally, the more precisely, than is less n_j. At real situations owing to trifle n_j equality (3.6) in general can be considered strict.

As for a pole order q^j, it is necessary to pay attention that it depends on concrete characteristics of structure most *j*-oh.

For further it is expedient to enter a number of designations:

$$(n_j)^{q_j} = b_j; \quad C_{-q_j}^j = a_j^2. \tag{3.7}$$

Taking into account the entered designations expression for the overall cost of all system of the protection consisting of *N* elements, which takes place at the considered moment, will have an appearance:

$$D = \sum_{j=1}^{N} f(b_j) = \sum_{j=1}^{N} \frac{a_j^2}{b_j}. \tag{3.8}$$

On the other hand, the received ratios allow approaching the solution of the task consisting in "distribution" of reliability between elements of structure for the purpose of ensuring the set reliability of P_Σ of all structure at the minimum financial costs of $D_{\min}$ of it.

Let's write down expression for total expenses of the structure providing its reliability:

$$D = \frac{a_N^2}{M - \sum_{j-1}^{N-1} b_j} + \sum_{j=1}^{N-1} \frac{a_j^2}{b_j}. \tag{3.9}$$

The following stage of the decision comes down to finding of the minimum value of expenses *of D*. For this purpose it is necessary to solve the equation $\frac{\partial D}{\partial b_j} = 0$ for $j = \overline{1, N-1}$.

Direct calculations lead to the following system of the equations:

$$\left(M - \sum_{k=1}^{N-1} b_k\right) a_j = a_N b_j. \tag{3.10}$$

The received system has the strict decision:

$$b_j = \frac{M a_j}{\sum_{k=1}^{N} a_k}. \tag{3.11}$$

Apparently, the required value of "reliability" is proportional to the corresponding coefficient a_j.

Coming back to initial parameters (3.7), the ratio (3.11) can be present in the form:

$$n_j = \sqrt[q_j]{\frac{\sqrt[M]{C^j_{-q_j}}}{\sum_{k=1}^{N} \sqrt{C^k_{-q_k}}}}. \tag{3.12}$$

At last, final expression for required probabilities will be define, though expression, bulky, but strict within the offered model:

$$P_j = 10^{-q_j \sqrt{\frac{\sqrt[M]{C^j_{-q_j}}}{\sum_{k=1}^{N} \sqrt{C^k_{-q_k}}}}}. \tag{3.13}$$

The final step by definition of the minimum expenses for ensuring the set probability of risk received, having substituted a ratio (3.13) in a formula (3.9):

$$D_{\min} = \frac{1}{M}\left(\sum_{j=1}^{N} a_j\right)^2 = \frac{1}{M}\left(\sum_{j=1}^{N} \sqrt{C^j_{-q_j}}\right)^2. \tag{3.14}$$

For assessment of efficiency of use of the allocated or available financial resource it is expedient to use as criterion the following natural relation standard in similar scientific tasks:

$$\mu = \frac{D - D_{\min}}{D} = 1 - \frac{D_{\min}}{D} = \frac{\Delta D}{D}. \tag{3.15}$$

Let's present the received ratio through the characteristics describing system elements, having used earlier received ratios:

$$\Delta D = \sum_{j=1}^{N} \frac{a_j^2}{b_j} - \frac{1}{M}\left(\sum_{j=1}^{N} a_j^2\right) = \sum_{j=1}^{N} \frac{a_j^2}{b_j} - \frac{\left(\sum_{j=1}^{N} a_j\right)^2}{\sum_{j=1}^{N} b_j}. \tag{3.16}$$

After simple, but bulky, transformations we will receive:

$$\Delta D = \frac{1}{2M} \sum_{k=1}^{N} \sum_{n=1}^{N} \frac{(a_k b_n - a_n b_k)^2}{b_k b_n}. \tag{3.17}$$

In this case parameter □ criterion will take a form:

$$\mu = \frac{\sum_{k=1}^{N} \sum_{n=1}^{N} \frac{(a_k b_n - a_n b_k)^2}{b_k b_n}}{\left(\sum_{j=1}^{N} a_j\right)^2}. \tag{3.18}$$

Let's consider a situation when in the investigation of those or organizational actions there was a change of cost of a certain element of system, respectively we will consider change of some a_j parameter for definiteness that this parameter is a_N. In this case, the question can be formulated as follows: "How it is necessary to change distribution of means to provide increase in reliability, or at former reliability to provide the greatest financial prize?".

Let's consider a situation, the general reliability of all structure remains invariable, that is $\sum_{j=1}^{N} b_j = M$, and we will define a prize which can be received. Let's replace a_N coefficient with new coefficient of c_N. Thus, we will have:

$$c_N = a_N - \varepsilon, \tag{3.19}$$

where □—change of coefficient a_N.

In case not to make any changes in distribution of reliability between elements of structure, then the financial prize which can be received, will be defined as:

$$\Delta D_1 = \frac{\varepsilon(2a_N - \varepsilon)}{b_N} = \frac{\varepsilon}{M a_N}(2a_N - \varepsilon) \sum_{j=1}^{N} a_j. \tag{3.20}$$

In case of redistribution of reliability between elements according to a formula (3.11) new values of coefficients b'_j will take a form:

$$\begin{cases} b'_j = \dfrac{Ma'_j}{\sum\limits_{k=1}^{N} a_k} = \dfrac{Ma_j}{c_N + \sum\limits_{k=1}^{N-1} a_k}; \quad j = \overline{1, N-1} \\ c_N = \dfrac{Ma'_N}{a'_N + \sum\limits_{k=1}^{N-1} a_k} \end{cases} . \tag{3.21}$$

The received ratios give the chance to learn how it is necessary to change the b_j parameters for maximizing reliability of all structure. If from ratio (3.21) subtract a equality (3.18), then we will receive:

$$\begin{aligned} \Delta b_j &= b'_j - b_j = \frac{\varepsilon M a_j}{\left(\sum\limits_{k=1}^{N} a_k\right)\left(c_N + \sum\limits_{k=1}^{N-1} a_k\right)} = \frac{\varepsilon b_j}{\sum\limits_{k=1}^{N} a_k - \varepsilon}: \\ \Delta b_N &= b'_N - b_N = \frac{\varepsilon (b_N - M)}{\sum\limits_{k=1}^{N} a_k - \varepsilon}. \quad j = \overline{1, N-1} \end{aligned} . \tag{3.22}$$

Apparently, reliability all N elements of system can be increased, therefore $\Delta b_N > 0, \quad j = \overline{1, N-1}$. Only the "fallen in price" structure where reliability generally can even be reduced is an exception $\Delta b_N < 0$.

The size of a prize can be found as a difference $\Delta D_{\min} = D_{\min} - D'_{\min}$ which will have in an explicit form an appearance:

$$\Delta S_{\min} = \frac{1}{M}\left[\left(\sum_{j=1}^{N} a_j\right)^2 - \left(\sum_{j=1}^{N} a'_j\right)^2\right] = \frac{\varepsilon}{M}\left(2\sum_{j=1}^{N} a_j - \varepsilon\right). \tag{3.23}$$

For final assessment of the turning-out prize in financial security it is necessary to subtract from the ratio determined by a formula (3.23) a ratio (3.20). As a result we will receive:

$$\Delta S_{\min} - \Delta S_1 = \frac{\varepsilon^2}{M}\left(\frac{\sum\limits_{k=1}^{N} a_k}{a_N} - 1\right) = \varepsilon^2\left(\frac{1}{b_N} - \frac{1}{M}\right). \tag{3.24}$$

The received ratio (3.24) gives the chance to estimate expediency of carrying out operations on redistribution of resources and reliability between elements of system of anti-terrorist safety.

3.2.2 Parallel Model

Let, as before, reliability *j*-oho of an element of system be characterized by probability of reliable work of P_{j0} where $j = \overline{1, N}$. Let also its successful functioning require some financial resource of d_{j0}. A natural way of increase in reliability of functioning of this element is connection of channels parallel to it. Let such channels there will be *n*. In this case the probability of reliable work *j*-oho of an element will increase and will be size:

$$P_j = 1 - \left(1 - P_{j0}\right)^n, \tag{3.25}$$

and ensuring its work requires a financial resource

$$d_{jn} = nd_{j0}. \tag{3.26}$$

Having defined value n from a ratio (3.25) and having set up him in (3.26), we will receive:

$$d_j = \frac{\lg\left(1 - P_j\right)}{\lg\left(1 - P_{j0}\right)}. \tag{3.27}$$

Apparently, the ratio (3.27) directly connects among themselves reliability of a new element with a financial resource.

Let's enter designation:

$$B_j = \frac{d_{0j}}{\lg\left(1 - P_j\right)}. \tag{3.28}$$

In this case functional communication between reliability of the system characterized by probability P_j, and a financial resource of d_j will have an appearance:

$$d_j = B_j \lg\left(1 - P_j\right). \tag{3.29}$$

The full resource providing the set reliability of system will have an appearance:

$$D = \sum_{j=1}^{N} B_j \lg\left(1 - P_j\right). \tag{3.30}$$

The ratio (3.30) represents logarithmic model of distribution of reliability at the set resource [1].

The presented model can be used for the solution of two optimizing tasks:

(1) How at the set probability of reliability of system $Q = \prod_{j=1}^{N} P_j$ to provide achievement it at minimum $D = D_{\min}$ resource? [1].
(2) How at the set resource D to provide the maximum reliability $Q = Q_{\max}$?

Minimization of a financial resource

Let's consider the solution of the first of the listed tasks: "How at the set probability of reliability of system $Q = \prod_{j=1}^{N} P_j$ to provide achievement it at minimum the $D = D_{\min}$ resource?" [1].

The requirement $\frac{\partial D}{\partial P_j} = 0$, where $j = \overline{1, N-1}$ brings to the following system of the equations:

$$\frac{B_j}{1-P_j} = \frac{Q}{P_j} \cdot \frac{B_N}{\prod_{i=1}^{N-1} P_i - Q}. \tag{3.31}$$

After elementary transformations, we will receive:

$$\frac{B_j P_j}{1-P_j} = \frac{Q B_N}{\prod_{i=1}^{N-1} P_i - Q} / \tag{3.32}$$

Considering that $Q = \prod_{j=1}^{N} P_j$, the received ratio can be presented in a symmetric form [1]:

$$\frac{B_j P_j}{1-P_j} = \frac{B_N P_N}{1-P_N}; \quad j = \overline{1, N-1}. \tag{3.33}$$

From the system of the equations (3.33) it is easy to receive:

$$P_j = \frac{B_N P_N}{B_N P_N + B_j(1-P_N)}; \quad j = \overline{1, N-1}. \tag{3.34}$$

Having multiplied among themselves all $(N-1)$ equations and having increased them by P_N we will receive [1]:

$$P_N \cdot \prod_{j=1}^{N-1} P_j = \frac{B_N P_N}{B_N P_N + B_j(1-P_N)}; \quad j = \overline{1, N-1}. \tag{3.35}$$

As P_N is close to 1, equality (3.35) can be presented in the form:

$$Q = P_N \prod_{j=1}^{N-1} \left(1 - \frac{B_j(1 - P_N)}{B_N P_N}\right) = P_N \left(1 - \frac{1 - P_N}{B_N P_N} \sum_{j=1}^{N-1} B_j\right). \tag{3.36}$$

Thus, expression for required probability will have an appearance [1]:

$$P_N = 1 - \frac{B_N(1 - Q)}{\sum_{j=1}^{N} B_j}. \tag{3.37}$$

Expressions for P_j where $j = \overline{1, N-1}$, it can be received from a formula (3.34). It is possible to arrive in the same way as expression was received for P_N.

Direct comparison of the turning-out formulas shows their coincidence to the same accuracy with what the ratio (3.36) [1] carried out.

It gives the chance to use a symmetric form of record [1]:

$$P_j = 1 - \frac{B_j(1 - Q)}{\sum_{j=1}^{N} B_j}, \; j = \overline{1, N}. \tag{3.38}$$

To find required minimum value of a financial resource, it is necessary to substitute a ratio (3.38) in equality (3.30) therefore we receive [1]:

$$D = D_{\min} = \sum_{j=1}^{N} B_j \lg\left[\frac{B_j(1 - Q)}{\sum_{i=1}^{N} B_i}\right] \tag{3.39}$$

It is expedient to efficiency of the offered distribution of the resource determined by a ratio (3.39) to estimate by comparison it with a case of uniform distribution between each of N elements [1].

In the latter case $P_j = \sqrt[N]{Q}$, and therefore according to a formula (3.30) the required resource will be equal [1]:

$$D_0 = \sum_{j=1}^{N} P_j \cdot \lg\left(1 - \sqrt[N]{Q}\right). \tag{3.40}$$

The efficiency of use of a resource can be estimate the relation [1]:

$$\eta = \frac{D_0}{D_{\min}} = \frac{\sum_{j=1}^{N} P_j \cdot \lg\left(1 - \sqrt[N]{O}\right)}{\sum_{j=1}^{N} \left\{ P_j \lg\left[\frac{(1-Q)P_j}{\sum_{i=1}^{N} B_i}\right]\right\}}. \tag{3.41}$$

Maximizing reliability of system

The essence of the second optimizing task consists in distributing the available financial resource D between elements systems so that reliability of system was maximum, so the value of probability of Q would be maximum.

Let's receive in an explicit form dependence of P_j from d_j that it is not difficult to make by means of a ratio (3.29):

$$P_j = 1 - 10^{d_j/B_j}. \tag{3.42}$$

Thus, the work is subject to maximizing:

$$Q = \prod_{j=1}^{N} \left(1 - 10^{d_j/B_j}\right), \tag{3.43}$$

under a natural condition that

$$D = \sum_{i=1}^{N} d_i. \tag{3.44}$$

Having carried out logarithmic differentiation of expression (3.43) on all variables d_j, and $j = \overline{1, N-1}$, and having considered that $d_N = D - \sum_{j=1}^{N-1} d_j$, we will receive the following system of the equations relatively d_j:

$$\frac{1}{B_j} \cdot \frac{10^{d_j/B_j}}{1 - 10^{d_j/B_j}} = \frac{10^{d_N/B_N}}{1 - 10^{d_N/B_N}} \cdot \frac{1}{B_N}; \quad j = \overline{1, N-1}. \tag{3.45}$$

The decision of system (3.45) passed approximately, its final look:

$$d_j = D \frac{\left[B_j\left(1 + \sum_{i=1}^{N} \lg B_i\right) - \lg B_j \sum_{i=1}^{N} B_i\right]}{\sum_{j=1}^{N} B_j}; \quad j = \overline{1, N}. \tag{3.46}$$

For determination of required maximum value of reliability, it is necessary to substitute the found expression (3.46) in a formula (3.43) therefore we will receive:

$$Q_{\max} = \prod_{j=1}^{N} \left(1 - 10^{d_j/B_j}\right). \tag{3.47}$$

As for efficiency assessment, it can is received when comparing with a situation of uniform distribution of a resource among all N elements, means at $d_j = D/N$, for which

$$Q_0 = \prod_{j=1}^{N} \left(1 - 10^{D/B_j N}\right). \tag{3.48}$$

Thus, the coefficient efficiency for the considered task will have an appearance:

$$\eta = \frac{Q_{\max}}{Q_0} = \prod_{j=1}^{N} \frac{1 - 10^{D\left(B_j + B_j \sum_{i=1}^{N} \lg B_i - \lg B_j \sum_{i=1}^{N} B_i\right)/\left(B_j \sum_{i=1}^{N} B_i\right)}}{1 - 10^{\frac{D}{d_j N}}}. \tag{3.49}$$

3.3 Statistical Models of Distribution of a Financial Resource Between Highly Reliable Elements of System. Model of "System of Ensuring Anti-terrorist Safety of Airline" and Model of Assessment of Probability of Realization of Terrorist Attack

It is possible to offer that in most cases distribution of a financial resource between elements of complex structure is made enough randomly "on an intuition" or by "the residual principle" that is most characteristic of the organizations with strictly limited budget and also on established to traditions and, as a rule, without any orientation to optimization by this or that criterion. Not always, it is obviously possible to carry out creation of analytical models as functional dependence of output parameters on input parameters, in certain cases it will be also insufficiently informatively. In these cases, it is possible to use statistical modeling by the classical Monte Carlo method [2].

In the considered class of applied tasks, in the general concept of ensuring anti-terrorist safety in airline, such approach threatens with excessively high costs connected with probability of penetration, considerable for such tasks, into conditional airline of the forbidden objects and substances, threats.

The optimizing task for the considered case was earlier already repeatedly formulated—at the set *financial resource* to make its such distribution between elements of structure to provide its *maximum reliability*.

On the other hand, it is possible to formulate also other optimizing task which will be acceptable in case reliability of system meets requirements of its "owner". Similar formulation of the question most is suitable for the solution of a problem of compliance of system of anti-terrorist safety of airline to conditions of an element of standard and legal regulation in the person of which the state and the law, the international normative legal acts which it demands strict observance acts (act as that requirement of "owner").

In this case, the concretized formulation of an optimizing task has the following appearance: "To provide the set level of reliability of system with the minimum financial resource".

Higher at a research all initial necessary ratios for the solution of the formulated task were received.

Influence of reliability on a resource as it made higher, we will describe by means of dependence like Loranov's number with a pole in a point, the corresponding reliability equal 1 or logarithmic dependence with a pole in the same point.

We will describe efficiency of use of a resource density of distribution of the relation of an initial resource to minimum required, that is $D/D_{\min}$ or the related size $\mu = (D - D_{\min})/(D + D_{\min})$.

Receiving analytical ratios in an explicit form is hardly possible. In this regard, the applied problem of statistical modeling, relying on generation of the random numbers D distributed on the set law put in the forefront. As such law, we will choose the uniform law as the most natural to the considered task.

Earlier for Loranov's approximation analytical expressions of population mean and dispersion of size μ of two-element system were received, that is it is applicable to a case $N = 2$.

For creation of required dependences in a case $N > 2$ the Monte Carlo method was used.

For $N = 4, 6, 8, 10$ are created random numbers in the range of 1 … 10. It is possible to assume that numbers in the specified interval cover a set of really arising situations. Further such numbers b_j are generated, that $\sum_{j=1}^{N} b_j = M$. Leaning on the received set of numbers, are calculated D and $D_{\min}$ also on the formulas received earlier:

$$D = \sum_{j=1}^{N} \frac{a_j^2}{b_j} \text{ and } D_{\min} = \frac{1}{m}\left(\sum_{j=1}^{N} a_j\right)^2, \tag{3.50}$$

where

a_j —the parameter characterizing a resource,
b_j —the parameter characterizing reliability.

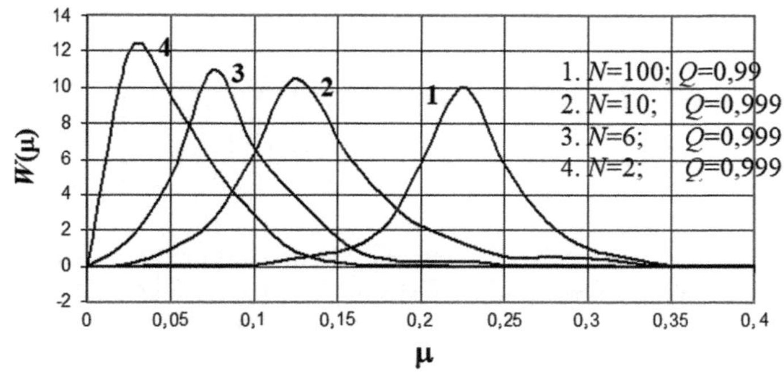

Schedule 3.3 Density of distribution of probabilities $W(\mu)$

Further it is made calculations of required effectiveness ratio:

$$\mu = \frac{D - D_{\min}}{D + D_{\min}}. \tag{3.51}$$

Such procedure repeats 10,000 times. Determination of density of probabilities $W(\mu)$, the giving opportunity finding of numerical characteristics—population mean and dispersion of a random variable μ. The received results are presented on Schedule 3.3.

For a special case of logarithmic approximation, assessment of efficiency of distribution of a financial resource between elements of highly reliable system of anti-terrorist safety of airline can be executed by means of the direct comparison which is available with the minimum resource $D_{\min}$:

$$D_{\min} = \sum_{j=1}^{N} d_j \lg\left[\frac{d_j(1-Q)}{\sum\limits_{k=1}^{N} d_k}\right], \tag{3.52}$$

where

d_j —the parameter characterizing a resource.
Q —the parameter characterizing reliability with expenses at uniform distribution of the D_0 between each of N elements.

$$D_0 = \sum_{j=1}^{N} d_j \lg\left(1 - \sqrt[N]{Q}\right) = \lg\left(1 - \sqrt[N]{Q}\right)\sum_{j=1}^{N} d_j. \tag{3.53}$$

Respectively the coefficient characterizing efficiency of distribution of a resource can be presented in the following form:

$$\mu = \frac{D_0}{D_{\min}} = \lg\left(1 - \sqrt[N]{Q}\right)\frac{\sum_{j=1}^{N} d_j}{\sum_{j=1}^{N}\left\{d_j \lg\left[\frac{d_j(1-Q)}{\sum_{i=1}^{N} d_i}\right]\right\}}. \tag{3.54}$$

For interpretation of the received results the Monte Carlo method made finding of density of distribution of probabilities of size μ, for what created numbers d_j from an interval 1 … 15 for numbers $N = 5; 10; 15; 100$ and $Q = 1 - 10^{-6}; 1 - 10^{-7}; 1 - 10^{-8}$.

On Schedule 3.4 density of distribution of probabilities are visually represented $W(\mu)$ at $1 - Q = 1 - 10^{-7}$ and various number of elements of system N_i. Specified demonstrates that with growth N there is a normalization of laws. As well as it was necessary to assume, increase in quantity of elements can lead to a big prize.

It is necessary to pay attention that on the schedule distribution of probabilities is shown further at $1 - Q = 10^{-8}$. At the same time, above and below-mentioned Schedules 3.4 and 3.5 are similar to extent of mixture, are identical that confirms weak dependence of density on number Q.

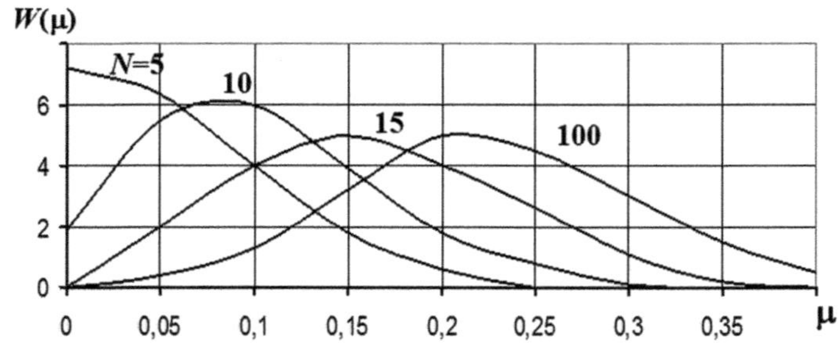

Schedule 3.4 Density of distribution of probabilities of $W(\mu)$

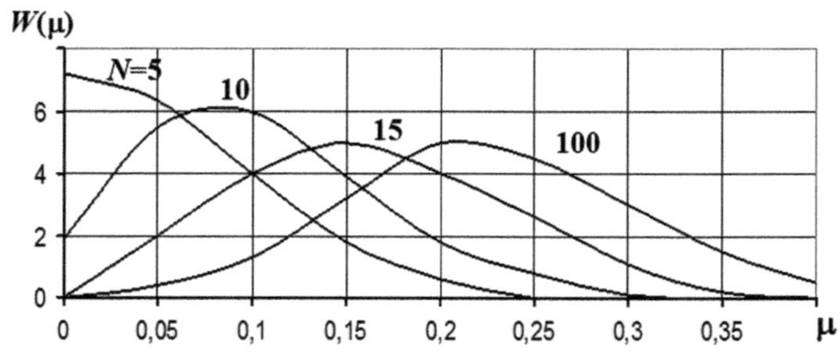

Schedule 3.5 Density of distribution of probabilities of $W(\mu)$

Model of "system of ensuring anti-terrorist safety of airline" and model of assessment of probability of realization of terrorist attack

As the main characteristic estimating the level of protection of a security system of airline against terrorist threat it is expedient to use *probability* (*risk*) that the system of anti-terrorist safety "will resist" such influence, that is its *reliability—P.*

This *probability* depends on perfection *of system*—a set of parameters, including its technical characteristics, a design, indicators of convenience of use, reliability, and so forth, at the same time the system consists of components—the structure of system joins all list of technical means of safety and other *i-elements*, which were in detail described at the beginning of the book. These characteristics making system elements, eventually, in absolute majority of cases defined by the system cost, some financial equivalent—*S.*

The security system of airline always consists of a set of elements among which all existing types of technical means of safety and also a human resource which acts as an independent element of protection (physical, barriers, customs, etc.) and as the operator of difficult security and control systems are combined (in aspect of organizational processes). For simplification of a conceptual framework, we will call further all set of such systems "*the system of ensuring anti-terrorist safety of airline*".

Within creation of system there are two approaches:

The first approach comes down to a problem of distribution of the available financial resource of *S* between systems for the purpose of maximizing reliability of all system—P_Σ.

The second approach proceeds from the requirement of minimization of total financial expenses at set (top in relation protected by system hierarchy level) the level of reliability P_0.

At the solution of the formulated higher than two optimizing tasks the problem of justification of functional dependence is central $P = f(S)$.

First of all we will pay attention that $P = f(0) = 0$, and at $S \to \infty$ value $P \to 1$. As in the real research, the aim of obtaining concrete numerical values not pursued, and the applied scientific task of identification of the general regularities and tendencies for decrease in risk of realization of terrorist attacks in airline is set, it is expedient to rely on the most widespread functional dependences of exponential and hyperbolic type.

In the first case we will have:

$$P = 1 - e^{-\alpha S}. \tag{3.55}$$

"Efficiency" of use of a financial resource in this case can defined evidently:

$$\frac{dS}{dP} = \frac{1}{\alpha(1-p)}. \tag{3.56}$$

Schedule 3.6 Functional dependences exponential (3.55) and hyperbolic (3.57) types

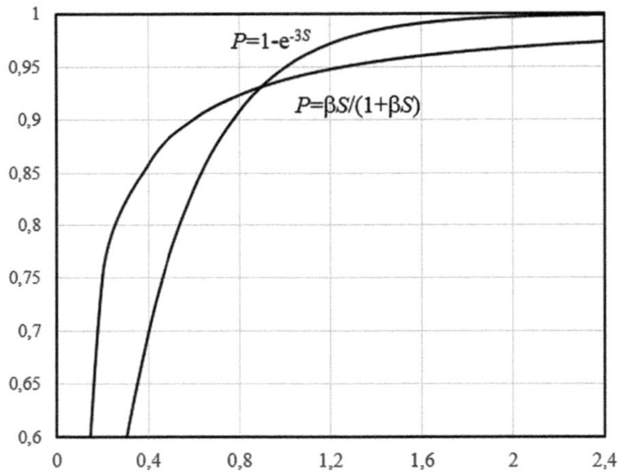

The given ratio demonstrates enough with what "work" in financial expression growth of reliability of system, so, for example, increase in reliability from value 10^{-6} to value 10^{-7} demands growth of expenses by ten times!

In case of hyperbolic approximation instead of Formula (3.55), we will have

$$P = \frac{\beta\, S}{1 + \beta\, S}. \tag{3.57}$$

Visually functional dependences exponential (3.55) and hyperbolic (3.57) types are shown in comparison on Schedule 3.6.

"Efficiency" of use of financial means in case of hyperbolic approximation will be expressed by a ratio:

$$\frac{dS}{dP} = \frac{1}{\beta(1-p)^2}. \tag{3.58}$$

Apparently, and in this case growth of reliability is followed considerable increase in necessary financial means. Let's consider consistently formulated higher than two optimizing tasks.

The system of ensuring anti-terrorist safety of airline represents set consistently connected N elements, reliability of each of which matters P_i ($i = 1, 2, \ldots, N$). To maintain reliability at the specified level, a financial resource in the amount of S_i. To maintain the required level of reliability during the operation of the system, a financial resource has been allocated in the amount of S_Σ. It is required to redistribute so a resource between elements to provide the maximum value of reliability of all system.

If it is about consistently connected N elements, then the probability of "failure"—P_{fail}, at least in one of elements, will be obviously equal:

$$P_{fail} = 1 - \prod_{i=1}^{N} (1 - P_i). \tag{3.59}$$

Using one of ratios (3.55) or (3.57), we will receive:

$$P_{fail} = 1 - \prod_{i=1}^{N} (1 - P_i(S_i)). \tag{3.60}$$

The equation imposing restriction for distribution of resources has an obvious appearance:

$$S_\Sigma = \sum_{i=1}^{N} S_i. \tag{3.61}$$

The presented set of the equations represents full system from the N-linear equations giving the chance to receive required ratios.

By way of illustration, the received ratios we will carry out detailed calculation for the system consisting of 2 elements.

System from 2 elements (hyperbolic approximation).

First optimizing task

In this case we have:

$$\begin{cases} P_1 = \frac{\beta_1 S_1}{1+\beta_1 S_1} \\ P_2 = \frac{\beta_2 (S_\Sigma - S_1)}{1+\beta_2 (S_\Sigma - S_1)} \end{cases} \tag{3.62}$$

The probability of failure will register in the case under consideration according to (3.60) as follows:

$$P_{fail} = 1 - \left(\frac{1}{1+\beta_1 S_1}\right)\left(\frac{1}{1+\beta_2 (S_\Sigma - S_1)}\right). \tag{3.63}$$

Having executed differentiation, we will receive:

$$\frac{dP_{fail}}{dS_1} = -\frac{\beta_1}{1+\beta_1 S_1} + \frac{\beta_2}{1+\beta_2 (S_\Sigma - S_1)} = 0. \tag{3.64}$$

From equality (3.64) taking into account (3.61) elementary, we find the solution of the first optimizing task for a case with two elements:

$$\begin{cases} S_{1opt} = \frac{\beta_1 \beta_2 S_\Sigma + \beta_1 - \beta_2}{2\beta_1 \beta_2} \\ S_{2opt} = \frac{\beta_1 \beta_2 S_\Sigma - \beta_1 + \beta_2}{2\beta_1 \beta_2} \end{cases} . \tag{3.65}$$

Second optimizing task

In the case under consideration, the reliability of the P_Σ system is set, in which it is required to distribute the reliability between the elements in such a way that the costs for this are minimal.

From a ratio (3.59) we will receive:

$$P_\Sigma = P_1 + P_2 - P_1 P_2, \tag{3.66}$$

from where follows

$$P_2 = \frac{P_\Sigma - P_1}{1 - P_1}. \tag{3.67}$$

Having used ratios (3.57), (3.62) and (3.67), we will receive:

$$\begin{cases} S_1 = \frac{P_1}{\beta_1(1-P_1)} \\ S_2 = \frac{P_2}{\beta_2(1-P_2)} = \frac{P_\Sigma - P_1}{\beta_1\beta_2(1-P_\Sigma)} \end{cases}. \tag{3.68}$$

Thus, the sum of full expenses of S_Σ can be present in the form:

$$S_1 + S_2 = \frac{P_1}{\beta_1(1 - P_1)} + \frac{P_\Sigma - P_1}{\beta_1\beta_2(1 - P_\Sigma)}. \tag{3.69}$$

For finding of the minimum S_Σ value we will execute differentiation of a formula (3.69) and we will equate the received expression to zero:

$$\frac{dS_\Sigma}{dP_1} = \frac{d(S_1 + S_2)}{dP_1} = \frac{1}{\beta_1(1 - P_1)^2} - \frac{1}{\beta_1\beta_2(1 - P_\Sigma)} = 0. \tag{3.70}$$

From a ratio (3.70) we find value of probability P_1 at which the minimum expenses will be provided:

$$P_1 = 1 - \sqrt{\beta_2(1 - P_\Sigma)}. \tag{3.71}$$

Respectively:

$$P_2 = 1 - \sqrt{\beta_1(1 - P_\Sigma)}. \tag{3.72}$$

Using the received ratios, it is easy to receive required value of the minimum expenses providing a preset value of reliability P_Σ:

$$S_{\Sigma\,\min} = \frac{\beta_1\left[1 - \sqrt{\beta_2(1 - P_\Sigma)}\right] + \beta_2\left[1 - \sqrt{\beta_1(1 - P_\Sigma)}\right]}{\beta_1\beta_2\sqrt{\beta_1\beta_2}(1 - P_\Sigma)} \tag{3.73}$$

The optimizing tasks and their decisions given in this paragraph demonstrated the importance of each of N-elements of system of ensuring anti-terrorist safety of airline into which structure enters as well a human resource. At the same time, as well as it stated above, the system can consist of a set of elements which main efficiency—in value reliability, as a rule is expressed in a financial equivalent that is especially important at the solution of the tasks of distribution of a resource given above in a research. Also the dependence of growth of reliability of system on a financial resource is visually traced: it agrees the acceptable level of reliability (risk) accepted by ICAO in aircraft estimated in 1×10^{-6}, growth of value 10^{-6} to value 10^{-7} will demand growth of expenses by ten times that objectively can be not always feasible in realities of modern airline at the current difficult economic situation.

3.4 Conclusion of the Chapter

1. The developed mathematical models allowed solving the following optimizing problems:
   - to provide the set reliability of system of anti-terrorist safety of airline at the minimum financial security;
   - to provide the maximum reliability of system of anti-terrorist safety of airline at the set financial security (minimization of a financial resource and maximizing reliability).
2. The developed models of statistical modeling allow estimating the probability of penetration of the terrorist on the protected object (territory) of airline from the number of the boundaries of protection having various levels of a burglary resistance to threat—reliability.
3. The constructed model of system of ensuring anti-terrorist safety of the airline consisting of a set of N-elements (subsystems) demonstrated the dependence of growth of reliability of the reliability on financial expenses: it agrees with the acceptable level of reliability (risk) accepted by ICAO in aircraft estimated in 1×10^{-6}, growth of value 10^{-6} value 10^{-7} will demand growth of expenses by ten times that objectively can be not always feasible in realities of modern airline at the current difficult world economic situation.

The practical importance of the results received in Chap. 2 is that the developed mathematical models allow increasing system effectiveness of anti-terrorist security of activity at airline in the way most of which is suitable for the concrete air enterprise, taking into account the set financial resource and requirements to reliability.

References

1. Hachikyan PP (2019) Problems of ensuring aviation safety at terrorist threats. The international research magazine, No. 12-1(90), pp 109–113
2. csc.sibsutis.ru [URL]: Methods of statistical modeling (Monte Carlo method). https://csc.sibsutis.ru/sites/csc.sibsutis.ru/files/courses/pvt/%20моделирование.pdf. Accessed 8 Dec 2019

Chapter 4
Mathematico-Psychological Models and Methods of Increase in Efficiency of Selection of the Experts Involved in Ensuring Anti-terrorist Safety of Airline

4.1 Requirements to Experts

The human resource, personnel—as an element of the system of anti-terrorist safety of airlines needs to be considered in aspect of the duties assigned to concrete service, departmental accessory, taking into account qualification and personal qualities, certain psychophysiological characteristics of the personality. It is necessary to define, by analogy with highly reliable technical means of safety that failure (refusal) of the conditional N-element "the identity of the employee involved in ensuring anti-terrorist safety" systems—can lead to terrorist attack commission that predetermines this element special, paramount importance.

The perspective of studying the identity of the employee involved in ensuring anti-terrorist safety is that it represents the difficult mechanism of the interconnected elements constructed both on personal and on especially professional, qualification qualities of the specific employee that, in general, allows the person to carry out the most responsible and decisive amount of works in this field of activity; at the same time, it is important to note that its replacement by a certain automatic system for the present is not represented possible. Also, scientists in the field of aircraft adhere to the similar point of view—specify V. V. and Mozolyako E. V. Sparrows "... such advantages of the person as ability to intuitive decisions in difficult in advance unforeseen situations still long time will not be able to be replaced in acceptable degree with automatic systems" [1].

The characteristic of reliability of the system of anti-terrorist safety is inseparably linked with such indicators as qualification and personal qualities of the experts involved in ensuring anti-terrorist safety. Reliability indicators j of system, i.e. abilities of these or those security service specialists to reveal the terrorist can characterize probability P_j of the fact that the terrorist will be beforehand revealed when holding these or those customs and control procedures, as will allow to prevent terrorist attack. Respectively, we will designate the probability that the terrorist will be able freely to pass j a boundary of protection as $(1 - P_j)$ (Scheme 4.1).

P. P. Hachikyan, *Countering Air Terrorism*, Springer Aerospace Technology,
https://doi.org/10.1007/978-981-19-6172-4_4

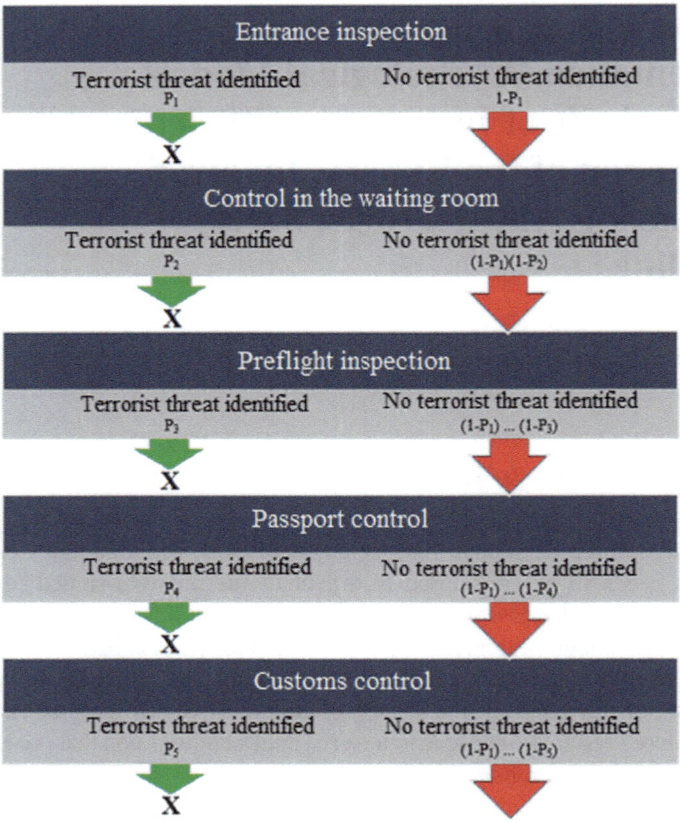

Scheme 4.1 Probability of identification of terrorist threat when passing various boundaries of protection of airline

Probability that the terrorist will make successful penetration with the forbidden objects or substances into the protected zone of airline because he passed all boundaries of protection, we will express as P_{Σ}:

$$P_{\Sigma} = \prod_{j=1}^{5} \left(1 - P_j\right). \tag{4.1}$$

Respectively, we will designate the probability that of P_0, that it will not be able successfully to pass all boundaries of protection as

$$P_0 = 1 - \prod_{j=1}^{5} \left(1 - P_j\right). \tag{4.2}$$

It should be noted that the specified probabilities of P_j belong to a situation of strict execution of the office duty regulations regulating safety in airline. Nevertheless, execution of similar instructions and accurately established requirements is implemented live, in every sense of this word, by the person. Despite its departmental accessory, qualification, and personal qualities, objectively there is a probability of appropriate execution of instructions depending on these or those internal and external factors, casual circumstances in concrete time point, in a stressful situation. Specified allows probabilities of P_j to give sense as instructor probability—$P_{j\,instructor}$, and to enter concepts of the real actual probability of execution by the official of the functions—$P_{j\,executor}$.

Thus, reliability j of the control system can be designated in the following look:

$$P_j = P_{j\,instructor} P_{j\,executor}. \tag{4.3}$$

In that case, the probability of penetration of the terrorist through j the control system should be written down as

$$P_0 = 1 - P_{j\,instructor} P_{j\,executor}. \tag{4.4}$$

Objectively, probability $P_{j\,instructor}$, as a rule, considerably surpasses value 0.9 therefore the paramount importance is gained by the value of probability $P_{j\,executor}$, which, depending on this or that the situation connected with the specific expert ensuring safety, can change in the range of values from 1 to 0.

The given example demonstrates the importance of both qualification and personal qualities of the employee involved in ensuring anti-terrorist safety. At the same time, its reaction in stressful situations which reasons can be a considerable list of factors, among which is important as preliminary estimate of personal and qualification qualities of the employee in the field of safety, and: discontent with the salary, tendency to corruption, family disorders, problems with health and different office conflicts.

Given defines that studying the personality has to be carried out systemically, covers the wide range of criteria undoubtedly significant for the general assessment of suitability of the specific candidate. For the assessment of qualities of the personality carrying out in-depth psychophysiological studies and testings which importance, unfortunately, the great number of employers represented by airlines neglects is necessary, relying on the general positive biographic data of candidates and, more—on the existence of recommendations, guarantees. Despite the reference of studying the personality mostly to science psychology, there are methods of mathematical modeling of assessment of the personality and decision-making that does possible studying of this perspective through a prism of technical science that is relevant for the real research. Influence of a human factor on the safety of flights was studied in works of the Russian scientists earlier in the field of aircraft [2–6], whereas questions of the influence of the specified factor when holding the corresponding procedures in the sphere of ensuring anti-terrorist safety and examination, according to the author of the book, is studied insufficiently in detail.

4.2 Qualification and Personal Qualities of Experts

Selection of the candidates involved in providing aviation and, in particular, anti-terrorist safety defined by the requirements established by industry standards and legal documents. Consideration began with the international normative legal act, so Chapter 8 Doc 8973 "The guide to aviation safety" of ICAO set the general principles of selection of the shots involved in ensuring aviation safety [7]. Point 8.3.2.4. application of an appropriate and reasonable technique which has to allow to recruit (quote) "… the most suitable and capable personnel for performance of operations on safety of civil aviation by a fair and effective way …" [7] is regulated. At the same time, considering the lack of references to concrete techniques, their choice remains to the discretion of airlines.

The main restrictive measures applied to persons by the connected ensuring transport safety are specified in article 10 of the Federal law of 09.02.2007 N 16-FZ "About transport safety" and have rather generalized character (existence of not removed criminal record, certain diseases, and so forth) [8].

More concretized restrictions and requirements are established by chapter 9 of the Federal system of ensuring aviation safety (The national program of aviation safety) in the general selection criteria of personnel.

At the same time, according to the same chapter of the normative document, in service of aviation safety the persons having standard negative characteristics such as are not accepted: the existence of unspent conviction, tendency to the use of drugs dismissed from state agencies for the negative reasons.

In point 215, it is noted that the structure of groups of preflight examination, has to include the persons having knowledge at least in one of the directions—psychology, engineer business, the operator of the X-ray equipment, the specialist cynologist, or having experience in civil aviation in the Federal Security Service, the Internal Affairs Service, border guard service, and customs authorities [9]. It is possible to note that existence in the structure of groups of preflight examination of the persons possessing according to point 215 knowledge in the field of psychology—just and assumes training of the corresponding persons in methods of profiling and will allow to approach a question of identification of potentially dangerous passengers from the point of view of psychology. The order of the Ministry of Transport of the Russian Federation of 21.08.2014 No. 231 also establishes a number of requirements to knowledge, abilities, skills of forces of ensuring transport safety, to personal (psychophysiological) qualities, level of physical training of separate categories of forces of ensuring transport safety [2]. This document is of interest the fact that, in its concrete requirements, qualities of candidates and lack of risk factors, are in detail stated.

Requirements with categorization for groups of positions are established also in the Federal system of training of personnel in the field of aviation safety (The national training program in the field of aviation safety) [10]. Concrete requirements to candidates are provided in the document and also the mechanism of preparation, retraining, and professional development of personnel is rather in detail stated.

It is also necessary to select the qualification requirements stated in the draft of the order of the professional standard of the Ministry of Labour and Social Protection of the Russian Federation—the expert of transport and aviation safety on air transport of civil aviation [11]. The professional standard planned for enforcement in 2020 provides the list of biographical and qualification requirements to the experts involved in ensuring transport and aviation safety; at the same time, concrete functional labor duties in the field of safety are given.

If the above-stated requirements concern only the civilians involved in ensuring aviation safety—the service of aviation safety (SAS), then, in connection with similarity of functions and tasks—it is necessary to also consider the staff of Internal Affairs, troops of national guard, and the Federal Security Service who also undergo the corresponding selection at reception on service. In comparison with civil normative legal acts, a more concrete and extensive example of requirements to candidates who are subject to studying when carrying out psychological selection can be seen in point 6 of the departmental rules of the Department of Internal Affairs approved by the Resolution of the Government of the Russian Federation of 06.12.2012 N 1259 "About the approval of Rules of professional psychological selection on service in Internal Affairs Service of the Russian Federation" [12].

Point 7 of the specified rules marks out risk factors which are subject to identification during performing comprehensive examinations (Scheme 17) [12].

The procedure and special psychological selection in divisions of troops of national guard of Russia are established by Rosgvardiya's order of 16.11.2016 N 357 "About the approval of the Instruction about the organization and carrying out professional psychological selection in troops of national guard of the Russian Federation" [13]. Rosgvardiya's instruction is similar to the contents on the existing documents regulating the activity of the Internal Affairs Service and police that is objectively based on transfers of a part of functions of Rosgvardiya from the Ministry of Internal Affairs, at the same time because of similarity of instructions, according to the author, she does not demand detailed consideration within the real research.

Considering the existing requirements to candidates when carrying out psychological selection for the Internal Affairs Service and divisions of troops of national guard, it is also necessary to address the conditions determined by the departmental order of FSB of Russia of 13.04.2011 N 151 "About the approval of the Instruction on the organization and carrying out professional psychological selection in staff of Federal Security Service", appendix No. 1 to the Instruction defined the following general requirements imposed to the citizens enlisting in security service under the contract, federal public civil service, work or on training in the educational organizations [14].

The provided departmental requirements and also the main requirements to personnel of SAB, allow to judge their community, orientation in general on the identification of certain typical positive lines of the personality and applicable qualification qualities in the real research. It should be noted that, regarding the ability of opposition to terrorist threat, according to the author, the general aspects of personnel (including psychological) selection, the services of aviation safety shown to employees should not differ from requirements to the staff of the Internal Affairs

Service, national guard, and the Federal Security Service. To be necessary to also pay attention that the existing problems of corruption, shortage and low qualification, and personal indicators of candidates can be the characteristic of candidates of all presented law enforcement agencies. Now, there was a situation in which the shortage of staff of services (Department of Internal Affairs, FSB, Rosgvardiya) can reach 40% that nevertheless, is not always publicly advertised in the form of similar figures because free (vacant) positions regularly reduce for carrying out optimization that formally leads to decrease in shortage, at the same time obligations and employment for service is shifted to already working employees forced to process. Despite qualifying standards to the personnel who are directly involved in safety, there is also a perspective of carrying out selection which is expressed both in employment refusal fit candidates and in wrong acceptance for work (service) of unusable candidates. There is also a separate problem at which successfully last selection the candidate, is incapable to carry out the duties assigned to it and is subject to dismissal for negative motives, at the same time similar can be connected both with objective errors of personnel selection, and with professional deformation, not the ability to resist to which led the worker (employee) to such state. A number of scientific articles and books in which features of carrying out selection, ways, and separate problem aspects are also considered as devoted to questions of personnel selection in the Internal Affairs Service [15–19].

The above and also the objective reasons in the form of already conducted terrorist attacks which reason the human factor was allowed to draw a conclusion on need of search of solutions of a personnel question in which, according to the author, the main problem is the quality of the made personal and qualification selection, underestimation of the importance of such procedures by the employer, insufficient work on their studying and improvement, need of withdrawal from settled, but not always enough effective methods. Further, the research of personal and qualification qualities of the persons who are directly involved in ensuring anti-terrorist safety is given in work, shortcomings of the existing methods of selection of shots are allocated and also new scientific methods and approaches to selection of shots, as on the basis of personal, and qualification qualities of candidates are offered that, being the solution of a problem of providing with suitable shots, will allow to reduce risk of commission of terrorist attacks in airline, will lead to work achievement of the goal.

In the comparative analysis of the importance of personal and qualification qualities of the specific candidate for the validity of concrete work, the linear dependence is unambiguously seen—the indicators are higher than both personal, and qualification qualities—the higher the ability to cope with performance of objectives. It should be noted that the argument would be objectively doubtful that the expert with low indicators of personal qualities could have high qualifications. Irrespective of it, at the comparative correlation of importance and value of personal lines and qualification qualities—according to the author, will be rational to carry out the selection of candidates by the principle of "the best of the best" according to both characteristics that most is suitable for selection of shots on important boundaries of safety, such as counteraction to terrorism on which during the work, the only made employee mistake—can become the last.

Eventually, the achievement of result (selection of the most suitable candidate) does not matter whether these or those lines and qualities of the candidate are congenital or acquired if the similar candidate for the moment of carrying out selection is objectively allocated with exclusively positive characteristics. Nevertheless, an opportunity to track the dependence of influence of certain inherited qualities and features on the personality which could be also put in the childhood and youth on further professional activity is rather interesting. So, do not raise doubts, the child from a dysfunctional family, who was brought up in an adverse environment, is doubtful that will be able successfully to work at the positions connected with law-enforcement activity further. For carrying out a comprehensive investigation of the identity of the candidate, it is necessary to decide on his qualities in projections to functions and all-social behavior, as will allow to use, further, these data at making decision on its employment (service). Questioning of the candidate for the purpose of clarification of his social environment, features of development, the brightest impressions from life and also obtaining information about the family of the candidate, a social status from third party, conditionally independent sources—poll of neighbors and relatives is rather justified that should not be limited to the formal characteristic from house management or the district police officer and, now, it is successfully carried out only at the selection of shots to security service. Candidates for positions of the employees involved in ensuring anti-terrorist safety should be estimated on 19 qualities.

Let's consider the qualities listed on the scheme in detail:

1. Personal qualities, in turn, are a set of parameters, which concern directly the identity of the worker. Among them—moral, social, and communicative. Personal qualities predetermine understanding and office the candidate bad from good, dobra from evil. Personal qualities, in general, can include questions of moral stability and other qualities, but, can be also considered as the generalized group.
2. Moral stability (tendency to temptation, anti-corruption stability, etc.) is the indicator of the personality characterizing ability to correspond the standard by traditions and also about indestructible dogmas—the good and evil, to an image of behavior. According to the author, it is necessary to include the factors of temptation and anti-corruption stability relating works to a subject most of which strongly influence the official at execution of the functions by him in opposition with terrorism in the same characteristic. At the same time, the tendency to temptation and anti-corruption stability—are closely interconnected with other qualities—personal, the level of requirements, reputation, but are always directly connected with material (bribery scale—the bribe size).
3. The motivation to work is considered as the ability of the personality to induce itself to work and to show the active actions directed to the achievement of objectives and purposes. The motivation which is directly proceeding from the personality without motivation from the employer (in the form of stimulation—awards, awards and so forth)—is rather difficult process which origin can be

connected as with internal desire of achievement of certain success, situation and the status, and with social or family factors at which material aspects are not considered by an individual.

4. Abilities are the specific features of the personality allowing the personality to develop and carry out a certain activity. The question of the existence of certain abilities and their origin (congenital—genetic, acquired in childhood at the correct education and so forth) is debatable in philosophy and psychology a subject, nevertheless, the development of certain abilities is a vector of achievement of good results in the chosen field of activity. It should be noted that lack per se of special abilities in childhood will not always demonstrate the achievement of good results or lack of prospects—abilities can be revealed spontaneously, in absolutely unexpected sphere and can be developed at a rather mature age.
5. Level of requirements can be considered both now, and in the long-term human lives. The greatest relevance for the corruption relations based on the level of needs of the bribetaker is represented by situational circumstances occurring for it spontaneously, at a given time, at which the bribetaker is tempted with receiving "fast", "easy" money, mistakenly believing that the suddenness of this action will remain unnoticed for the controlling structures. In a case with situational approach which is relevant in this work, the level of requirements can be a basis for the development of the corruption relations between terrorists and the security service specialist. The famous American physiologist, A. Maslou, represented requirements—steps in the form of a pyramid with basic needs [20].

According to A. Maslou, these requirements predetermine behavioral model and adoption of certain decisions by the person. In a case with an assessment of needs of the official as corruption and inclined element, undoubtedly, physiological, and some existential requirements will prevail over the others.

1. Health—the general parameter defining a condition of the person. According to the definition by the world association of health care (WHO): "Health is a condition of full physical, mental and social well-being and not just absence of diseases or physical defects" [21]. At the same time, it should be noted that, in the context of this research, special listens deserves also mental and psychological health of the persons which are carried out by functions on the safety of airline. Besides, both the civil, and certified personnel, when carrying out employment (service) without fail undergo the procedure of professional selection—passing of medical commission (which objectively does not enter studying within the real research) and special psychophysiological researches.
2. Professional and business qualities are set of the parameters characterizing the level of professional competences of the specific worker as well acquired by it in the course of work, for example, the ability to carry out objectives in time, discipline, accuracy, etc.
3. Qualification needs to consider as the narrow parameter demonstrating the presence of an individual of a certain narrow knowledge and abilities in a specific industry. Qualification, as a rule, also distinguished by the existence of a certain

qualification category, a class or rank (in a case with different service in state agencies).

4. Productivity and effectiveness are indicators of the efficiency of work of an individual. The efficiency is a ratio of time spent on performance work with the received result.
5. Educational level should be considered the indicator demonstrating the presence of the person of a certain education level. In the Russian Federation, five education levels—preschool, initial, average, the highest, and postgraduate, are officially established.
6. Labor mobility defines the ability of the personality to adapt to changing labor duties. This includes readiness for the sake of labor prospects to master a new profession and specialization, to undergo professional development, and also in a literal sense to follow work (readiness for moving, change of the residence).
7. The reputation is considered as the fixed opinion on a concrete individual which is based on the actions which are earlier made by it in certain situations. High-quality performance of objectives, integrity, and firmness—shown to society (to a circle of people, collective, and so forth) at some point, undoubtedly, set persons positive reputation, at the same time level of the requirement of society of result from actions of the similar person—will always be at the high level that in turn generates at the person from high reputation—certain fears for the reputation and sets a tendency of continuous development (growth).
8. Knowledge—considered as the general parameter of cognitive activity of the person. In a widespread explanation for knowledge, there corresponds the concept of an outlook and outlook which systematize an overall picture of knowledge ordered by the person in his life.
9. The concept "ability" defined how a certain set of knowledge is put into action. The ability as the way mastered by the person, allows trying to obtain objectives, the base at the same time is always knowledge.
10. Experience is knowledge acquired in the course of activity. At the same time, it is necessary to include in the set of this definition as an individual experience, the data on experience of other people acquired directly in practical activities, so received at knowledge acquisition—the transferred experience. For example, mentoring is a striking example of the mechanism of the transferring experience.
11. The skill characterizes a possibility of the personality to perform a certain work and to achieve results with the special, high level of professionalism. As a rule, the skill, is the acquired property of the personality, fundamental which personal experience, knowledge is applied by the particular person in the best execution.

By consideration of qualities and lines of the personality, it is necessary to decide on the factors exerting impact on the identity of the employees involved in ensuring anti-terrorist safety as whom the author allocates also:

1. The factor of the environment includes the impact on the identity of society, the immediate environment and family. It is simpler to explain it as a factor of the habitat of the person in society (irrespectively the nature).

2. The factor of climate and food defines the impact on the identity of climatic conditions (nature) of the residence or work and connected from these inconveniences (or opposite) and also food as a component of the general standard of living.
3. The genetics as a factor of influence is rather a debatable question, nevertheless, it is worth believing that it is possible to call the specified factor and biological, having included in its process of the early birth and development of the child at even irresponsible age.
4. A case—that factor which is the most difficult for predicting and simulating as definitely the built sequence of events can represent such case in which the person develops a special idea of how it should come to a concrete period.

Despite the existence of techniques of ranging of the factors exerting impact on the employees involved in ensuring anti-terrorist safety [22], it should be noted that, in a case with the selection of the candidate for a responsible position—the current state of the candidate for the selection moment whereas its state has to be exposed further to assessment already in the course of work is decisive and deserves special attention of competent services of airline for control of personnel, psychological services, services of assessment of efficiency, etc.

Considering psychological aspects of the personality involved in ensuring anti-terrorist safety, it is also necessary to note such characteristics as tendency of the official to the corruption relations. Psychological tendency to corruption of officials can serve one of the ways of distribution of terrorism in aircraft. Corruption (from the Latin words "*corrumpere*"—to corrupt also "*corruptio*"—bribery, damage)—the settled term [23], designating use by the official of the powers of authority and entrusted it is right and also connected with this official status of its owner, opportunities and communications for personal benefit, benefit of other interested persons having a certain interest in permission of this or that situation and a controversial issue [23, 24]. Official legal definition of corruption is reflected in Art. 1 of the Federal law "About Anti-corruption": this abuse of official position, bribery, taking of a bribe, abuse of authority, commercial bribery, or other illegal use by the natural person of the official capacity contrary to legitimate interests of society and state for obtaining benefit in the form of money, values, other property or services of property character, other property rights for or for the third parties or illegal providing such benefit to the specified person by other natural persons and also commission of the acts stated above from a name or for the benefit of the legal entity [25]. The air enterprise providing transportation of passengers and freight owing to features of functioning is of special interest for the development of the corruption relations. For a reward—a bribe, to the potential terrorist, it becomes available: to carry out a bringing of the forbidden objects and substances aboard the aircraft, to get on the territory with limited access, and to carry out other actions connected with performance of problems of terrorist character. The corrupted employee cannot always realize that he becomes the direct helper and the collaborator of the act of terrorism. True intentions of the malefactor can be hidden from the bribetaker—for example, instead of a bringing rather "harmless" suitcase of not declared sanctioned products of food without appropriate examination, the malefactor can bring on board the aircraft the improvised explosive device. Undoubtedly, a similar lack of information in true intentions of the malefactor terrorist, in no way cannot justify actions of the corrupted security service specialist. It is also necessary to allocate the corruption phenomenon as general means of shadow financing of terrorism. The impressive income gained by the affiliated organizations of terrorists by means of corruption schemes with the assistance of officials of various levels—creates the enormous criminal budget for the preparation and commission of acts of terrorism worldwide. Corruption in society consists not just in the formal legal term and group of the criminal actions relating to it, and includes and something more extensive is a steady public phenomenon with the bright social, economic, and psychological components inherent in modern society. Considering questions of corruption, it is necessary to categorize the persons involved in safety in the isolated groups, which are created on belonging to a certain profession or departmental accessory.

Staff of service of aviation safety and the auxiliary private security companies and also persons, directly not involved in safety, are civil personnel that, in general, makes them rather free and a little controllable for the higher management, in a type of lack of tough standards (charters, obligations according to federal laws, and so forth). In opposition to it, staff of the Department of Internal Affairs, Rosgvardiya, and the staff of the border service (FSB) are the persons serving in staff of police, internal service, and military service, respectively, at the same time on the specified categories of persons departmental normative legal acts, such as Federal law "About Service in Internal Affairs Service of the Russian Federation" of 30.11.2011 N 342-FZ, the Federal law "About Police" from 07.02.2011 N 3-FZ, the Federal law of 03.07.2016 N 226-FZ "Expatiate on troops of national guard of the Russian Federation", and the Federal law of 04.05.2000 N 55-FZ "On the Border service of the Russian Federation", the Federal law "About the Status of the Military Personnel" of 27.05.1998 N 76-FZ that puts special obligations and requirements on specified persons. It is also necessary to note that the solution of a perspective of the absence of qualified personnel and due control among civil personnel, is the possibility of the creation of separate anti-terrorist group in the existing service of aviation safety of airline in which persons from among civil personnel of SAB can enter on a constant basis, the PSF which is most prepared and the counteractions to terrorism having experience where on non-staff basis experts from law enforcement agencies also without fail have to be attracted, at the same time, a task of the specified group will fight against only terrorism [26].

Unfortunately, the personnel work which is carried out at enrollment on service to the Internal Affairs Service, Rosgvardiya, and the border service of the Federal Security Service takes place rather superficially, without speaking already and the personnel selection made in SAB of airlines and various PSF. Despite the use of certain psychological techniques, tests for career guidance, and also passing of medical board—they carry rather all-exacting, in relation to a certain average norm, character, without profound studying of the personality and psychology of the specific candidate [23]. Only the obvious inadequate behavior of the candidate will be the cause for refusal at enrollment on service in similar organizations [23]. Questions of corruption, in relation to military service in the Russian Armed Forces (that, nevertheless—practically in fragmentary degree belongs also to military service in the staff of FSB of Russia) are in detail covered in the corresponding profile monograph by the author on jurisprudence [27]. The catalyst of the "simplified" reception on service is also staff shortage in divisions of law enforcement agencies at positions of younger command structure at which shortage of staff can reach 40%. A corruption problem as a way of distribution of terrorism in aircraft allocates as a basis of emergence of the similar criminal relations shortcomings of work of HR and psychological departments on which obvious defect, unreliable persons appear on service, at extremely responsible boundaries of safety.

4.3 The Analysis of the Existing Methods of Selection of Experts

Various methods of research of the personality applied at employment (service) in the state departments gained distribution and in recent years in the commercial sector which directly includes most the air enterprises. Carrying out researches is carried out on the basis of the medical and flight commissions of experts (MFCE) of civil aviation whose main objective is the complex psychophysiological diagnostics (check of health, including mental, and psychological) regarding the validity of pilots and aviation personnel, at inclusion in higher educational institutions of civil aviation in various directions of preparation; however, the specified specialties are not the faces which are directly ensuring safety from terrorist threats [28]. Within the real research, attention to the psycho diagnostics which is carried out at reception on service of the persons who are directly dealing with safety issues and being the certified personnel also deserves (police officers and internal service Department of Internal Affairs, Rosgvardiya, the military personnel). At reception on service, at transfer to other position, and under other circumstances established by the relevant departmental provisions, the listed persons undergo complex psychophysiological diagnostics. Holding procedures is regulated by both internal departmental documents, and the approved open orders—the Resolution of the Government of the Russian Federation of 06.12.2012 N 1259 "About the approval of Rules of professional psychological selection on service in Internal Affairs Service of the Russian Federation" [12] Order of FSB of Russia of 13.04.2011 N 151 "About the approval of the Instruction on the organization and carrying out professional psychological selection in staff of Federal Security Service" [14], Rosgvardiya's Order of 16.11.2016 N 357 "About the approval of the Instruction about the organization and carrying out professional psychological selection in troops of national guard of the Russian Federation" [13]. Techniques of testing change and improved, adapt under concrete department (service), can differ between various divisions and in territoriality, however, in general, consist of a complex enough widespread and long since applied by psychologists, among which: testing on SMIL L. Sobchik (in various variations—full from

566 statements, the reduced look—398 statements, without information on additional scales) [29], testing on MMIL F. Berezina and M. Miroshnikova [30], PITCHES—the reduced multiple-factor questionnaire for a research of the personality from 71 statements and 11 scales developed on the basis of the SMIL [31, 32], test and also tests of intelligence quotient (IQ) of G. Ayzenk [33], the CAT—the Short indicative, selection test of V. Buzin, E. Vanderlik, the color test of M. Lyusher, and others [34].

As the final stage of carrying out psycho diagnostics (as a rule, in case of end of I and II stages—successful passing of tests and the subsequent conversation with the psychologist) the technique of check of the examinee of the candidate on a polygraph was widely adopted. The polygraph represents the modern technical complex (as a rule, on the basis of the personal computer) intended for carrying out a tool psychophysiological research of parameters (cardiovascular activity, electric resistance of skin, pulse rate, and breath) is synchronized with a statement to the examinee of various questions. Check mechanism based on a polygraph is a comparison of the kept reaction (medical indicators) of the examinee's organism at the answer to test questions (answers to which are not subject to doubt in reliability) with the reaction of an organism to control questions (directly questions, answers to which are subject to check by the principle—a truth/lie). It is scientifically proved that the results of check on a polygraph have unreliable, unscientific character, and check of a large number of people showed results not better than casual guessing that demonstrates doubtfulness of the use of a similar system of check for the solution of responsible personnel tasks. Besides, the results received on a polygraph happen doubtful because of fear of the particular person of passing of check called by the extended and widely lit false beliefs (including from the staff of human resources departments, polygraph examiner, psychologists) about "force", "reliability" and "check depth" which, allegedly, can reveal the truth even in attempt of intended deception and concealment of the facts and circumstances. In confirmation of the position about insolvency of results of a printing research, it is necessary to allocate a position of the Supreme Court of the Russian Federation, which specified in the explanations—"... such researches aiming at development and verification of investigative versions do not treat proofs according to Art. 74 of the Code of Criminal Procedure of the Russian Federation" [35], and also practice of rejection of results of a research as proofs in the judicial system of the USA and other countries of the world [36].

On the basis of the above, it is obviously possible to draw conclusions on the expediency of carrying out the selection of candidates as on personal, and separately on qualification qualities, using various methods of poll and testing. The results of researches of candidates received by means of conducting check on a polygraph have a doubtful character that is confirmed by world practice in this connection application of a polygraph by the author of the real research is not recommended.

The techniques of a research of the personality applied constructed, as a rule, on the interpretation of results by experts by psychologists in a rather wide range, at the same time, making decision on the validity of the candidate to a concrete kind of activity is assigned only to a subjective discretion of the expert conducting a research and carries through character, recommendatory for the person of the making decision, nevertheless—as a rule, it is decisive. A similar approach inevitably leads to employment (service) of the candidates not capable to carry out the functions, and refusal in reception of fit candidates because of mistakes at the selection of shots (what more just to describe as a classical problem of the emergence of errors of the first and second sort). Further, it is offered to consider the existing techniques, to reveal their negative sides, and to offer the ways directed to the improvement of the procedure of selection of shots.

Let's consider the widely applied method of a research of the personality with MMPI abbreviation. In the 1930s, famous American erudite S. Hatuey and D. McKinley dealt with the issues of diagnostics of the personality by means of testings (polls). The personal questionnaire published in 1940 was called in honor of the university at which they also carried out the professional activity—the Minnesota multidimensional personal questionnaire (Minnesota Multiphasic Personality Inventory, abbr.—MMPI) [37]. became result of their long-term works. Initially, the development of the test was caused by need of carrying out the professional selection of pilots whose problem was separation of pathology from norm [37]. The basis of the test is a quantitative comparison of answers of representatives of standard groups to answers of patients (scientific statistical base) at whom certain frustration in extremes of their manifestations are revealed. Simplicity of conducting testing consists of the examinee in the choice of a statement from the test concerning itself by the principle "truly" and "is not true". In 1967, the MMPI test was adapted and processed by the Soviet scientists, F. Berezinymi and M. Miroshnikov, received the new name TMRP (Technique of a Multilateral Research of the Personality), made 384 statements for the examinee [30]. Now, the most demanded is the SMIL test from 566 statements under authorship L. Sobchik developed on the scientific base of the Leningrad psychoneurological institute of V. M. Bekhterev [29].

A scientific statistical base within the real work data of the SMIL system according to L. Sobchik on additional scales is accepted. It is also necessary to note that in general, application of the questionnaire of SMIL is rather demanded that is visually shown on the Scheme 4.2.

Testing technique basis on SMIL is the statistical base allowing differentiating norm from pathology. The test results received in the form of digital values are suitable for further mathematical analysis and combination with other systems.

In a usual look, the procedure of testing for SMIL is performed on the following algorithm:

1. Filling with the examinee of the questionnaire with the general data and making the written consent to conducting testing and data processing.
2. Definition by the psychologist on the basis of the filled-in questionnaire of gender accessory of the examinee, in connection with various questions for males and females.
3. Acquaintance of the examinee with the instruction for carrying out testing.
4. Conducting testing by filling with the examinee of the form of the test with marks about fidelity of a statement or record of results of answers in the special questionnaire, is also possible filling of the test in electronic form on the personal

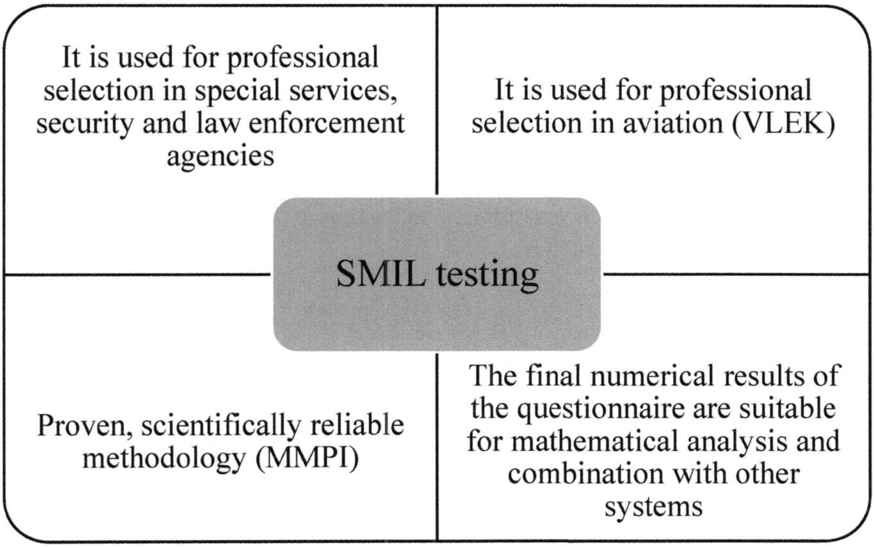

Scheme 4.2 Justification of the use of system of testing of SMIL with allocation of the main advantages

computer (depending on the level of material support of psychophysiological laboratory).

5. Completion of testing, transfer results of testing for processing and analysis to the psychologist.
6. Delivery and execution of the conclusion about psycho diagnostics concerning the examinee the psychologist (commission) on the basis of results of imposing of keys, interpretation of the test, and these other forms of studying of the personality (if it is provided)
7. In case it is provided by rules of the held testing—discussion of results of the examinee with the psychologist, issue of recommendations of need of further observation, the direction to profile experts, on treatment or additional studying of the personality, and so forth.

In standard option, the analysis of test results is made by means of interpretation of scales which are made on the basis of comparison of indicators of the examinee with the statistical data on patients having these or those concrete psychiatric and psychological diseases:

– *Scales of reliability of passing of the text (3 scales, estimated—L, F, K)*

Scale of L

The scale also carries the name of "a lie scale". The indicator of a scale shows desire of the subject in more "positive" light, to show strict observance of social norms which, actually, are not always possible for executing, certainly, if it does not become defiantly and on display. According to the author, despite the attempt of authors of the questionnaire to strictly determine by a similar scale desire of the subject to expose itself better than it is actually and due to this criterion to nullify the test passed by the subject (because of its not franknesses)—not always it will be true. In reality, high-organized persons with accurate, and sometimes both excessively strict moral and social foundations which, according to the author of the questionnaire, will look by results of this scale—false meet.

F scale

The scale also carries the name of "reliability". The raised profile on a scale demonstrates distortion by the subject of results of poll, at the same time, it is possible to define that data are distorted by it deliberately or accidentally only based on the ratio of this scale with the others, making the general analysis of results.

Scale To

The scale also carries the name "corrections". A problem of a scale is to reveal persons who try to hide or reduce the appearance of some mental states during a research.

F-K index

It is also necessary to note such parameters as the index of scales of F and K. Then the difference F-K is more, especially the aspiration of the subject to emphasize weight of the symptoms and vital difficulties, to cause sympathy is expressed. In turn, the low index F-K reflects the aspiration of the subject to prove to be in more best light, to deny existence of certain problems.

Indicator (scale) "?"

In essence, it is impossible to call an indicator with the sign "?" a scale as it fixes only the number of statements of the subject on which that could not give the concrete answer. The indicator will matter if exceeds 26 crude points as figure 26 corresponds to the number of the statements withdrawn from shortchanging described in the instruction to the test questionnaire as: Number of this statement should be led round a circle". If the indicator of a scale "?" is higher than 70 crude points, data of the test will be considered as doubtful. The general figure within 36–40 crude points is also admissible and results from 41 to 60 points already demonstrate excessive vigilance of the subject. The main ten scales (clinical):

- *Additional scales*

For years of existence of techniques of MMPI and SMIL, authors of tests questionnaires and also the practicing psychologists, depending on the relevance of these or those questions, created a set of additional scales for carrying out in-depth studies [29]. Additional scales have a special importance within the real research as allowed to mark out the most important lines of the personality which are directly exerting the impact on performance of functions by it in the candidate as served as their acceptance as statistical scientific base.

For carrying out the real research by the author the existing additional scales given in Appendix A to the book are used.

Despite broad application, the standard option of application of a technique of MMPI or SMIL has a number of shortcomings. According to the author, on the basis of these techniques it is necessary to carry to objective shortcomings of selection of candidates:

1. Wide range of norms under which the value of results on scales from 30 to 70 (75) points is accepted. At the same time, the specified interval of norms is treated by the author of the questionnaire only on the basis of statistical data (that it is possible to believe, it is based on a certain universality of the test, but not its specificity for selection of particular persons with determined by qualities). Despite it, according to the author of the book, the specified interval is too wide that is especially unacceptable in the selection of shots on responsible boundaries of safety on which as was stated above—it is supposed to make selection of candidates by the principle of their greatest compliance by the certain criteria which are listed above in the present chapter.
2. Lack of an opportunity to give at the end of testing a definite answer about the validity of the candidate for a concrete position (it is suitable or it is not suitable), whereas the results received as a result of testing have extremely ample opportunities for their interpretation by the specialist psychologist (in case of lack of an extreme of value of results). Existence of a subjective factor in the person of the psychologist is a lack of similar testing at which the suitable candidate can be unreasonably rejected (is not recommended or conditionally recommended) only because the specific expert who was incorrectly interpreting the results of testing so solved. Besides, the basis of similar conclusions of the psychologist can be not always properly finished or commented in an available form to the person (to the head of HR department and so forth), directly employing decision on the reception of the candidate (service) in this connection this person should be trusted only qualification of the psychologist and absence at it the subjective reasons in discredit of the specific candidate.
3. The prevalence and common knowledge of software testing to techniques of MMPI and SMIL to which passing the candidate can be prepared and develop strategy of answers to questions so that to give results in the range of norm (which as it was already stated in item 1 above—has rather wide range) that will lead to the emergence of a mistake at selection (employment (service) of the actual improper candidate).

The listed shortcomings allow to formulate problem definition in the field of improvement of psychological professional selection of candidates as follows:

(a) to change a norm interval when carrying out the selection of shots and to make it adjustable upon the demand of the highest step of hierarchy of personnel selection (the head of the HR department, the top officials of the employer, in the long term the relevant normative legal act and so forth) for improvement of quality of selection of candidates at whom on service (work) the most suitable will get on personal qualities. At the same time, to allocate with the right of change of an interval (reduction, increase) exclusively top level of hierarchy—heads of the organizations, HR departments, having limited the admission to the regulation of level of the persons which are directly carrying out the selection on places (ordinary personnel workers, psychologists and so forth);

(b) to apply less widespread—new (and the hierarchies of personnel selection which are variably changing by means of minor changes of the top step) test techniques (to which successful passing according to hints by providing doubtful data on the personality the candidate will not be able (or it, at least, will be more difficult) to be prepared previously);

(c) to reduce the influence of a subjective factor in the interpretation of results, at the same time as a method the mathematical model based on the principles of classical Boolean algebra capable after data input of answers of the candidate to issue the unambiguous conclusion about its validity can be applied.

For improvement of psychological professional selection, considering the above shortcomings of a technique of SMIL (MMPI), according to the author it is necessary to consider the possibility of application of the theory of mathematical psychology. Works of the famous Soviet and American psychologist-mathematician V. A. Lefebvre are devoted to a combination of psychology and mathematics in questions of behavior of the person. Evidence-based connection of psychology with mathematics allowed to look at the most difficult questions of free will, belief, and moral of the person with absolutely new, excellent from widespread, the points of view. In the numerous works, V. A. Lefebvre was the first in the USSR who developed the theory of reflexive games and innovative approach in works on mathematical modeling of conscience in correlation to concepts "well" and "badly", "yes" and "no", without participation of ethics, having set calculation of these sizes on the basis of classical Boolean algebra. In the scientific article, the member correspondent of RAS D. A. Novikov characterizes the importance of the works of V. A. Lefebvre in the following words: "V. A. Lefebvre's works on reflection models, on reflexive games and the monograph—made, in a sense, a scientific revolution, having given to wide scientific community (mathematicians, engineers, sociologists, economists, etc.) language and the device of the constructive analysis of reflexive processes, and the monograph became well-known and axiomatic …. each of the mentioned scientific directions actively develops so far, D. Harshanya and R. Aumann got Nobel Prizes, the corresponding results are included into classical textbooks on game theory, but the tradition to refer to V. A. Lefebvre's works, unfortunately, is absent" [38]. According to the author, questions of free will and mathematical modeling of behavior of possible acts of the security service specialist of the air enterprise are most important for the solution of personnel tasks in the selection of employees whose zone of responsibility includes a decrease in a possibility of commission of acts of terrorism in aircraft. The special attention, within the real research, is deserved by the scientific works of V. A. Lefebvre devoted to "a formula of the person" in which acts of the person are interpreted by figures "1" and "0", also, as well as the world around pressure upon it is described on the similar poles exerting impact on it (positive—"1" or negative—"0"). The similar mathematical apparatus allowed to receive in the form of results simple values which can be used for a combination of results with other systems most fully suitable for application in the real research.

4.4 Probabilistic Method of Selection of Experts in Qualification Qualities

The main criterion for making decision on a possibility of appointment of the person to a position of the employee involved in ensuring anti-terrorist safety is, eventually, the amount (number) of successful results, when carrying out over it qualification tests. Similar tests can have character of the control questions and tasks set to the subject, or they can have estimated character of behavior of the person in various naturally or artificially created situations and serve as assessment of his qualification. At last, they can be directly connected with congenital psychological characteristics of the person, with his educational and cultural level that also reflects its qualification. However, in all cases, in making decision on an opportunity to use the candidate for the performance of this or that work, the person making decision—the head, the psychologist, and the employee of the HR department, estimating results of tests at the final stage, comes up against a dilemma, and in fact against a situation, binary, characteristic of Boolean algebra,—he has to give the concretized answer: "is suitable" or "it is not suitable", "1" or "0", respectively.

We formalize the considered situation with the help of mathematical approach. In relation to the considered position to which the employee is accepted, his assessment is carried out on the basis of the certain qualities reflecting educational, vocational, and so forth characteristics. Let the total number of such characteristics will be N. It means that it is possible to speak about some N-dimensional space of signs where each candidate for a position A can be characterized by the corresponding N-dimensional vector And.

Here, it is necessary to pay attention to the fact of dependence or independence among themselves of such characteristics. At independence—it will be a question of orthogonal vector space. In case of dependence among themselves of several or all characteristics—it will be a question of affine space which can be orthogonalized that gives the grounds for all further reasonings to carry to orthogonal space. For convenience, we will call the orthogonal characteristics further corresponding "qualification characteristics". The length of a vector has to act as the natural numerical characteristic A,

$$\text{i.e. } |A| = \sqrt{\sum_{i=1}^{N} a_i^2}. \tag{4.5}$$

It is natural that any of the used qualification characteristics, that is ai, is not described by a continuous function. The discrete description of values is for this purpose used. For example, all axis a_i breaks on discrete $a_{i1}, a_{i2} = a_{i1} + \delta$ and so on where δ—discrete.

Thus, the N-dimensional space of signs breaks into N^N of pieces of N-dimensional "cubes" in one of which there "is" a candidate for "position".

From the point of view of "employer", the situation when characteristics of "candidate" in accuracy get to the required qualification N—measured "cube" will be ideal. In such approach, *the N-dimensional* distance between this "cube" and "cube" to which the candidate belongs can serve as a natural measure of compliance of "candidate" of held "position". In case of the different importance (weight coefficient) of qualification signs as natural selection criterion of candidates, the mentioned distance taking into account the weight importance of coordinates will still act. It is important to emphasize that the decision, eventually, made on *one* number.

At selection of candidates for similar responsible positions, the emergence of errors of two types is inevitable: employment of the candidate who is not conforming to requirements, and, on the contrary—refusal in reception to the candidate meeting relevant requirements. It is obvious that both such decisions have a casual character and therefore it is expedient to speak about the probability of adoption of this or that decision. Let's designate the probability of *wrong reception*—p_1, and the probability of *wrong refusal* in reception—p_2. In this case, assessment of the *single decision* can be characterized as the natural sum:

$$S = s_1 p_1 + s_2 p_2.$$

In case of "*mass*" selection of candidates, it is expedient to use for decision-making widespread "Neumann-Pearson's criterion" in the radio world [39], at which the probability of false alarm is fixed (wrong reception—p_1) and the probability of the correct decision is maximized (the probability of refusal in reception—is minimized—$(1 - p_2)$).

The essence of tests comes down to obtaining answers to the asked questions, and then calculation of the correct and wrong answers on comparison of these numbers is made the decision is made. In relation to the situation considered in the real research, it is not about the number of answers and about the number of the qualification signs carrying the candidate for a position to number suitable.

In relation to the considered case affirmative answer is a compliance to this or that qualification requirement (coordinate in *N-dimensional* space). To define compliance of the candidate—it is necessary to ask the examinee *exactly N* questions (to carry out N qualification tests). However in practice to the examinee significantly less questions—n, that is $n < N$, and sometimes even $n \ll N$. Naturally, at the same time, emergence of mistakes about which it was told above, *wrong refusal*—p_1 and *wrong reception*—p_2.

Thus, the applied scientific task is set as follows:

The candidate for a position is objectively characterized by N independent qualification qualities.

From the employer's position, candidates break into a quantity of categories for which number we will designate *M letter*. By the first highest qualification category, we will carry candidates and amount of necessary qualification signs n which satisfy the inequality: $N < n < N_1$. Similarly, for the subsequent categories: $N_1 < n < N_2$, $N_2 < n < N_3$, etc. Thus, each candidate objectively has this or that category.

When conducting tests the candidate is checked how it was already told above, on a significantly smaller number of qualification signs which we will designate a letter $Q < N$.

For making decision on the determination of category of the candidate, the employer also needs to break Q scale into the same number of categories M.

Thus, objectively, there is the following applied scientific task: "What probability that the candidate by results of tests referred to category *i*, objectively belongs to the category *j*?".

This probability we will designate as p_{ij}.

The second circle of tasks is connected with a question as it is necessary to break the Q scale to provide the smallest probabilities of p_{ij} within the marked-out criterion.

For further reasonings, it is expedient to present a set of probabilities p_{ij} in the matrix form. Without limiting the community to further consideration, we will be limited to the case shown in Scheme 4.3, that is, we will speak about five categories.

We will present a required matrix of probabilities in the form, and we will call it a matrix of mix-up of hypotheses:

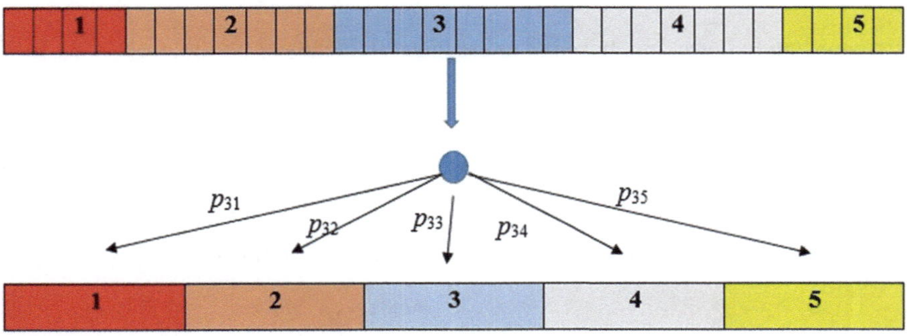

Scheme 4.3 Schematically representation of reference of the candidate to categories

$$P = \begin{pmatrix} p_{11} & p_{12} & p_{13} & p_{14} & p_{15} \\ p_{21} & p_{22} & p_{23} & p_{24} & p_{25} \\ p_{31} & p_{32} & p_{33} & p_{34} & p_{35} \\ p_{41} & p_{42} & p_{43} & p_{44} & p_{45} \\ p_{51} & p_{52} & p_{53} & p_{54} & p_{55} \end{pmatrix}$$

From the given record, it is visible what takes place in 5 situations relating to authentic decisions and 20 to wrong.

If to enter the cost of each decision, then in a general view instead of a matrix P it is necessary to use the S cost matrix:

$$\mathrm{S} = \begin{pmatrix} s_{11}p_{11} & s_{12}p_{12} & s_{13}p_{13} & s_{14}p_{14} & s_{15}p_{15} \\ s_{21}p_{21} & s_{22}p_{22} & s_{23}p_{23} & s_{24}p_{24} & s_{25}p_{25} \\ s_{31}p_{31} & s_{32}p_{32} & s_{33}p_{33} & s_{34}p_{34} & s_{35}p_{35} \\ s_{41}p_{41} & s_{42}p_{42} & s_{43}p_{43} & s_{44}p_{44} & s_{45}p_{45} \\ s_{51}p_{51} & s_{52}p_{52} & s_{53}p_{53} & s_{54}p_{54} & s_{55}p_{55} \end{pmatrix}.$$

The given matrix gives the chance to estimate the cost of wrong decisions and to make the relevant administrative decision. The quality of the made administrative decisions can also be subjected to assessment, proceeding from criteria of safety, feasibility, etc., [40] nevertheless, within the research existence of the decision which does not demand a similar estimation is supposed.

The correct decision gives a prize which it is possible to estimate expression:

$$S_+ = \sum_{j=1}^{M} p_{jj} S_{jj}. \tag{4.6}$$

According to costs, it is possible to estimate by means of a ratio:

$$S_- = \sum_{\substack{i,j=1 \\ i \neq j}}^{M} p_{ij} S_{ij}. \tag{4.7}$$

If to enter concepts of balance, then expression for it will have an appearance

$$S_{общ} = S_+ - S_- = \sum_{j=1}^{M} p_{jj} S_{jj} - \sum_{\substack{i,j=1 \\ i \neq j}}^{M} p_{ij} S_{ij}. \tag{4.8}$$

The ratio (4.8) defines the damage estimated in some conditional sizes which gives the chance to approach some quantitative assessment of opportunities for the candidate. The question of numerical value of the given ratios of probabilities appears in is central.

We will begin consideration with an alternative situation. In this case, it is about a bi-variant situation. After certification on m from M of qualification signs, it is necessary to take out the conclusion about a possibility of use of the specific candidate.

The employer considers that if the number of qualification signs of the candidate of l from the total number of signs of L is more than some threshold value $l > l_0$, then the candidate objectively belongs to the first category and can be employed. Otherwise, the situation is opposite.

When checking candidates, $Q < L$ signs are subjected to control. The threshold q_0, value defining limit of categories is here too entered and happens identification of objective first category to the category appropriated to it, kind of conditional, category. At such situation, as already earlier it told, there are errors of two types caused by a mix-up of the decision on the assignment of categories.

Let's carry out the calculation of probabilities of these events.

The matrix of mix-up of hypotheses will have dimension 2 × 2 and will take the following form:

$$P = \begin{pmatrix} p_{11} & p_{12} \\ p_{21} & p_{22} \end{pmatrix}.$$

The probability that among L selected qualification features, exactly l_0 features will belong to the first category:

$$P_L^{l_1} = \frac{C_{Q-q_0}^{L-l_1} C_{q_0}^{l_1}}{C_Q^L} = \frac{q_0!L!(Q-q_0)!(Q-L)!}{Q!l_1!(Q-q_0-L+l_1)!(q_0-l_1)!(L-l_1)!}, \tag{4.9}$$

where $C_a^b = \frac{a!}{b!(a-b)!}$—binomial coefficient.

The candidate will conform to the requirements of the employer if the number of its qualification characteristics on $l_1 \geq l_0$ or on $(l_1 + 1)$ the characteristics, or on $(l_1 + 2)$ characteristics, etc., up to the L characteristics.

At last, probability p_{11} the fact that the candidate who is objectively corresponding to the first category will quite so it is estimated as a result of test two alternative tests can it is found by the corresponding summation of expression:

$$P_{11} = \sum_{l_1=l_0}^{M} \frac{C_{Q-q_1}^{L-l_1} C_{q_1}^{l_1}}{C_Q^L} = \sum_{l_1=l_0}^{M} \frac{q_1!L!(Q-q_1)!(Q-L)!}{l_1!Q!(Q-q_1-L+l_1)!(q_1-l_1)!(Q-l_1)!} \tag{4.10}$$

Having a ratio (4.10), there is a probability p_{12}:

$$P_{12} = 1 - P_{11} = \frac{1}{C_Q^L} \sum_{l_1=0}^{l_0-1} C_{Q-q_1}^{L-l_1} C_{q_1}^{l_1}. \tag{4.11}$$

The probabilities connected by wrong assessment of the candidate are similar. It means that the number of qualification characteristics of the candidate of l are less than l_0, that is, $l_1 < l_0$.

Told gives the chance by analogy to make an expression for probability that the candidate who is objectively not belonging to the first category will be mistakenly referred to this category of candidates:

$$P_{21} = \frac{1}{C_Q^L} \sum_{l_1=l_0}^{L} C_{Q-q_1}^{L-l_1} C_{q_1}^{l_1} \tag{4.12}$$

and, respectively,

$$P_{22} = 1 - P_{22} = \frac{1}{C_Q^L} \sum_{l_1=0}^{l_0-1} C_{q_2}^{l_1} C_{Q-q_2}^{L-l_1}. \tag{4.13}$$

Formulas (4.9)–(4.13) define a required matrix of mix-up of hypotheses for the considered case.

Let's illustrate the received ratios. The question of the dependence of probability of the correct classification of the candidate on the number of the compared qualification characteristics is central.

The probability that exactly l of classification characteristics of the candidate from their total number L belongs to the candidate can be calculated by a formula:

$$P(Q, L; l) = \frac{C_L^l C_{L-l}^{Q-L}}{C_L^Q}. \tag{4.14}$$

The schedules provided further give an evident idea of the nature of influence of the number of the analyzed qualification characteristics on required probability (Schedules 4.1, 4.2, 4.3, 4.4, 4.5, 4.6, 4.7, and 4.8) for employees of 3–4 categories.

The provided schedules give the chance to visually estimate the influence of the number of the checked qualification signs on the probability of the correct answers

Schedule 4.1 Probability of receiving exactly m of the correct answers from employees of category 1 ($Q = 25$), category 2 ($Q = 50$), category 3 ($Q = 75$), category 4 ($Q = 90$)

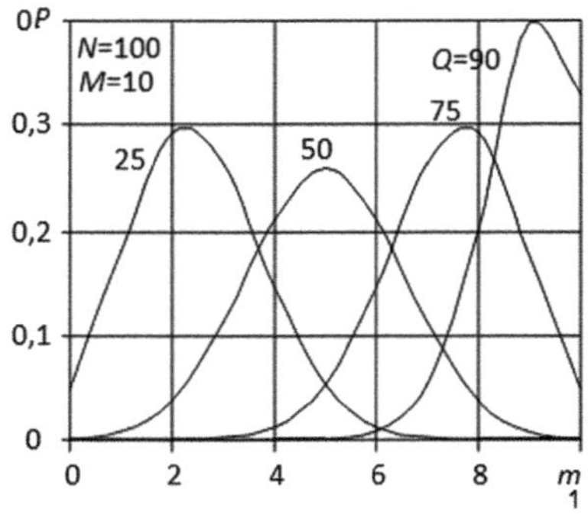

Schedule 4.2 Probability of receiving exactly m of the correct answers from employees of category 1 ($Q = 10$), category 2 ($Q = 25$), category 3 ($Q = 40$)

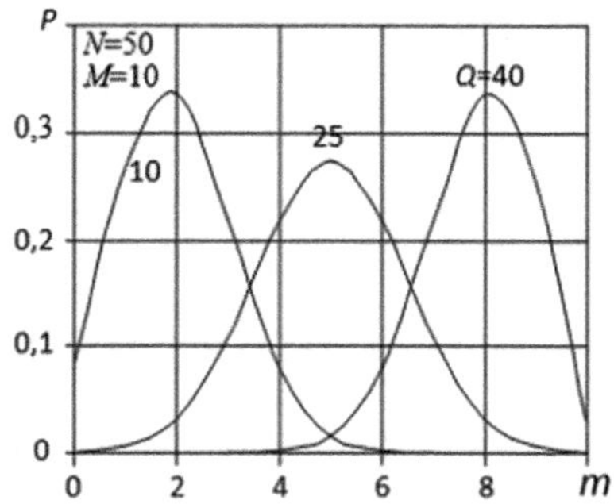

Schedule 4.3 Probability of receiving exactly m of the correct answers from employees of category 1 ($Q = 4$), category 2 ($Q = 10$), category 3 ($Q = 16$)

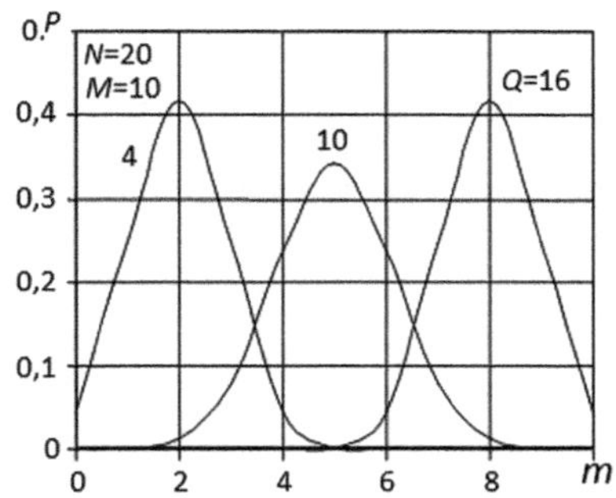

Schedule 4.4 Probability of receiving exactly m of the correct answers from employees of category 1 ($Q = 25$), category 2 ($Q = 50$), category 3 ($Q = 75$), category 4 ($Q = 90$)

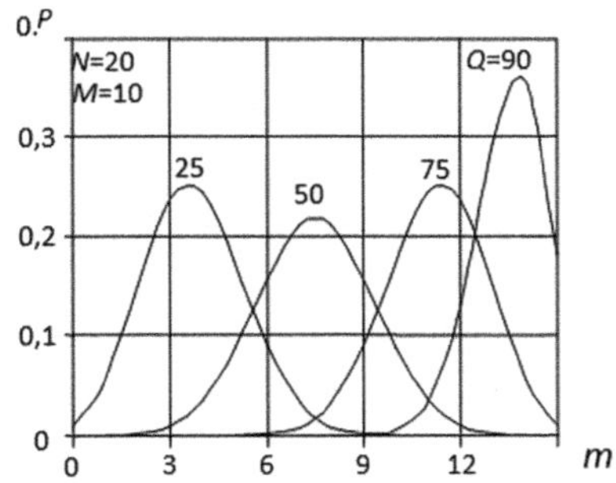

Schedule 4.5 Probability of receiving exactly m of the correct answers from employees of category 1 ($Q = 10$), category 2 ($Q = 25$), category 3 ($Q = 40$)

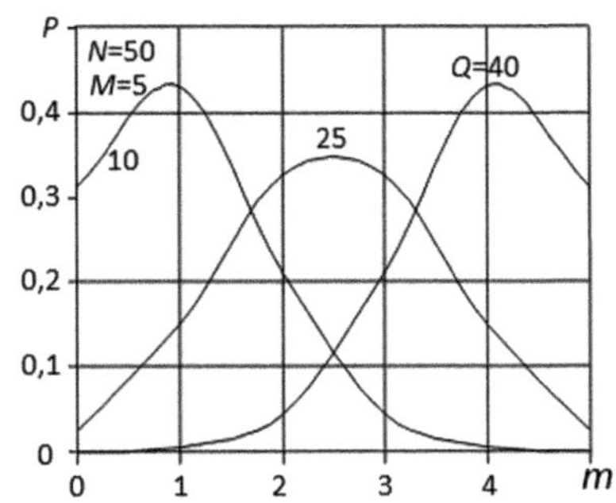

Schedule 4.6 Probability of receiving exactly m of the correct answers from employees of category 1 ($Q = 4$), category 2 ($Q = 10$), category 3 ($Q = 16$)

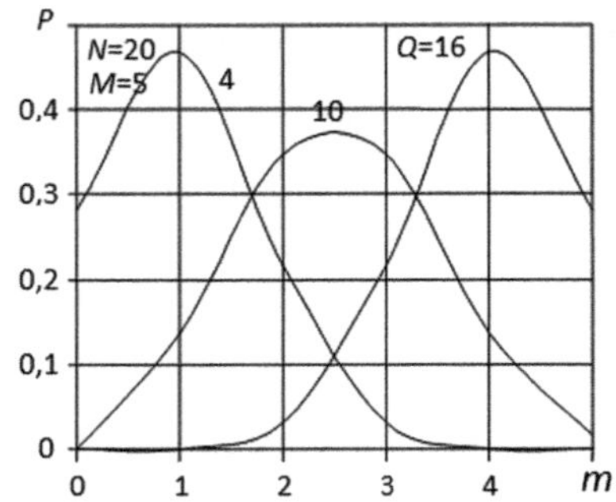

Schedule 4.7 Probability of receiving exactly m of the correct answers from employees of category 1 ($Q = 10$), category 2 ($Q = 15$), category 3 ($Q = 25$), category 4 ($Q = 40$)

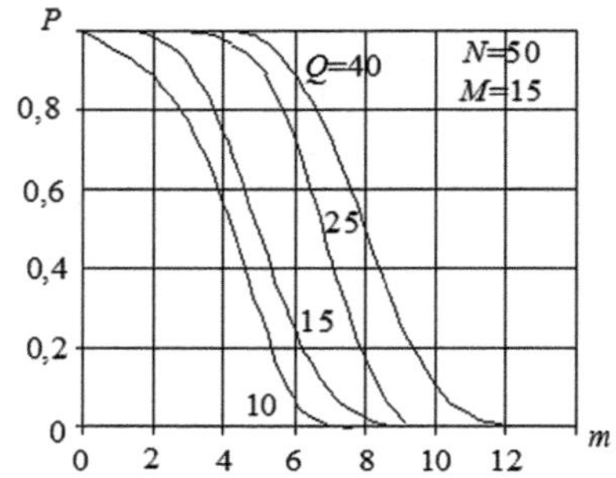

Schedule 4.8 Probability of receiving exactly m of the correct answers from employees of category 1 ($Q = 5$), category 2 ($Q = 10$), category 3 ($Q = 15$), category 4 ($Q = 25$)

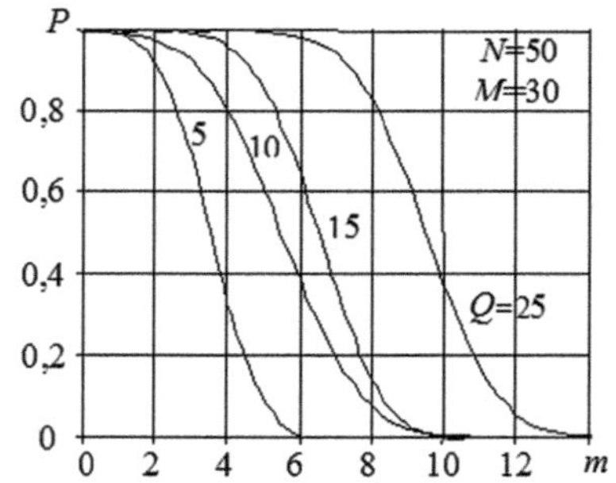

for candidates for positions of the employees involved in ensuring anti-terrorist safety with various qualification signs.

4.5 Methodical Recommendations About the Selection of Experts in Personal Qualities

Personal qualities of the expert involved in ensuring anti-terrorist safety allow it to make important decisions, at the same time in the depth of such decisions the foundation free will which cannot be objectively analytically predicted with complete certainty is laid fundamental. There is a set of modern interpretation of a psychological phenomenon of free will of the person. The academician of RAS B. B. Kadomtsev believes that the free will is a freedom of action or the free choice, at the same time the person has free will and is free in commission of the acts, even if the person and is influenced by any circumstances—the choice all the same will remain behind it. The known populizer of science S. Hawking in the works considered that actions of people set by the certain physical system operating all real, and the impossibility to predict the choice is determined by an element of chance that in fact and is the declaration of free will of the person [41]. But, there is also a mathematical option of an explanation of fundamental bases of adoption of decisions by the person that can be seen in V. A. Lefebvre's works [42, 43] adapt and is applicable key bases of the theory of the famous scientist to scope of the real author's research:

1. If the employee who is responsible for anti-terrorist safety cannot present himself the helper of terrorists or otherwise the allowing illegal act, then at any external impact on it from outside irrespective of—whether there will be this influence in a negative or positive side, this influence will be turned into real action of the employee.
2. If the employee who is responsible for anti-terrorist safety can present himself the helper of terrorists or the different way making an illegal act, then he is capable to refuse commission of a similar act.

Similar configuration of statements displayed mathematically as follows:

1. Let's enter variables a and b.
2. We will take for a what the employee really tests and wishes to make (*simply—"I this way will not act"*), for b—acts which are made by its subconscious image of itself (*simply—"I believe that I so would not (arrive)"*).
3. Let's define variables on a set $\{1, 0\}$, considering that "*0*" value—"an illegal act", and "*1*"—"a lawful act".

Widespread it is possible to call such situation when ensuring anti-terrorist safety of airline when the employee makes an illegal act, for example, getting a bribe for connivance at examination and realizes for himself that commission of a similar act for it is quite possible and admissible, in that case $a = 0$ and $b = 0$, respectively.

Let's set function $f(a, b)$, which will connect real acts and alleged actions of the candidate with his real actions. In that case, the statement above under the item 1 "*If the employee … This influence turned into real Action of the employee*" can be represented in the form of [43]:

$$f(a, 1) = a, \tag{4.15}$$

whereas a statement under the (4.16) *"If the Employee … Then He Is Capable to Refuse a Similar Act"* in a look:

$$f(a,0) = 1. \quad (4.16)$$

The similar design can be presented in the form of a binary logical connective:

$$f(a,b) = b \rightarrow a \quad (4.17)$$

The general concept of similar interpretation of actions of the employee is based that certain internal deep experiences of the employee for the negative actions are directed to actually prevention of commission of similar illegal acts. The question of the nature of emergence of this judgment in habitual aspect relies on psychological and physiological features of brain activity of the person. Lefebvre in the works believes to find possible a ratio (4.17) a certain mathematical essence of the person by means of which is possible to open deep communications of the phenomenon of the person and his sensual and cogitative activity.

For creation of a mathematical possibility of commission by the employee responsible for ensuring anti-terrorist safety of airline, concrete acts, we will specify the following variables and we adapt their definitions [43]:

1. x_1—*a measure of pressure of the state and society declining the employee to commission of a lawful act*;
2. $1 - x_1$—*a measure of pressure of the state and society declining the employee to commission of an illegal act (respectively, it can be presented in the form* $(0 \le x_1 \le 1)$);
3. x_2—*value judgment by the employee of pressure of the state and society declining it to commission of a lawful act*;
4. $1 - x_2$—*value judgment by the employee of pressure of the state and society declining it to commission of an illegal act (or in a look* $(0 \le x_2 \le 1)$*)*;
5. x_3—*internal aspiration of the employee (intension) to make a lawful act*;
6. $1 - x_3$—*internal aspiration of the employee (intension) to make an illegal act (or in a look* $(0 \le x_3 \le 1)$*)*;
7. X_1—*a measure of readiness of the employee to make a lawful act*;
8. $1 - X_1$—*a measure of readiness of the employee to make an illegal act (or in a look* $(0 \le X_1 \le 1)$*)*;
9. p—*the employee as the formal operator acting on a set of couple* $(p,\ 1-p)$, *where* $(0 \le p \le 1)$.

It is necessary to consider process of influence of the employee (the formal operator) which in essence has to exert impact on x_3, $1 - x_3$ and translate them, respectively, in $X_1, 1-X_1$—that is in essence the operator will transform internal aspirations of the employee (intension) moderately readiness to make this or that act. As couple $(p,\ 1-p)$ the component can use p, that allows to present action of the employee in the form of the following formula:

$$Sub_{x_1x_2}(x_3) = X_1 \tag{4.18}$$

The similar treatment gives the chance to allocate for the employee involved in ensuring anti-terrorist safety, a mathematical type of the conscious choice which implies that its internal aspirations (intensions) are correlated to a measure of its readiness (that is commission of a lawful or illegal act is not the emotional or "sudden" choice) that it is necessary to present in the form $x_3 = X_1$. It also allows to draw a conclusion that Sub_{x_1,x_2} on an own element is a theoretical analog of the conscious choice.

Within the real research also that circumstance that in scientific works of V. A. Lefebvre the free will is determined by morals (does not depend on external influence—and is considered as quite idealized option) is interesting that is especially appropriate in questions of adoption of such important decisions which enter competence of the employee involved in ensuring anti-terrorist safety of airline. Lack of external influence—high-morality, it is possible to consider quite real in case personnel selection is made successfully and (service) such person who is capable to follow determined morals is really employed, external influence if is not excluded—that is minimized to the limits of an extreme close to 1. Such "idealized option" can be accepted as existence at least of one couple of values $x_1 = a$ and $x_2 = b$, at which any internal intension of the employee turns into readiness to make specific action for which the above-stated operator Sub_{x_1,x_2} is identical [43]:

$$Sub_{a,b}(x) = x \tag{4.19}$$

for any $x \in [0, 1]$, where $x = x_3 = X_1$.

On the other hand, there are always such external factors which can make impact on the security service specialist and entail change of its moral principles and foundations, to force to change own $x_3 = X_1$ under action $1 - x_1$, among which it is necessary to call:

1. Understanding of a real possibility of loss of own life (lack of resolute readiness on self-sacrifice for the sake of society, bigger number of people and so forth).
2. Capture by terrorists in hostages and/or existence of real threats of life and health to persons from among relatives and relatives of such employee for change of adoption of certain decisions by it.
3. The corruption interest of the employee differing in the size of a bribe, bribery.
4. Various jet mental states caused by the psy-injuring situation.

In similar cases when the employee makes lawful or illegal actions, but regardless of the intentions, it's necessary to accept as existence not less than two couples of such values $x_1 = c$, $x_2 = d$ и $x_1 = e$, $x_2 = \mathrm{f}$, that can be shown as [43]:

$$Sub_{c,d}(x_3) = 1, \tag{4.20}$$

$$Sub_{e,\mathrm{f}}(x_3) = 0 \tag{4.21}$$

for any $x_3 \in [0,\ 1]$.

From the specified statements and formulas (4.19, 4.20, 4.21) it is obviously possible to adapt three main axioms of the theory of V.A. Lefebvre as follows:

1. *The axiom of "trustfulness"* claims if the state and society are represented by the employee ideal ($x_1 = 0$) (from the subjective point of view of the employee, do not induce him to commission of illegal acts ($x_2 = 1$)) [43], then the employee responsible for ensuring anti-terrorist safety of airline is always ready to make any, including an illegal act upon the demand of the state and society.

$$Sub_{0,1}(x_3) = 0 \tag{4.22}$$

for any $x_3 \in [0,\ 1]$.

The similar behavior of the employee in aspect of perception of ideality of the world through trustfulness, most authentically reflects an incident picture when a terrorist attack is conducted without any financial investments and bribery, is exclusive from ideological (the personal reasons) or political reasons, on the basis of all-consuming belief (radical religious, extremist, etc.). The trustfulness and imposing of false ideals are used by ideological inspirers and recruiters of terrorists, substituting at the candidate for terrorists of idea of the ideal state and society, true universal values.

2. *The axiom "not evil intentions"* claims if the state and society motivate on compliance with law and about ($x_1 = 1$),then the employee responsible for ensuring anti-terrorist safety of airline anyway will make only a lawful act that can be shown in the following look [43]:

$$Sub_{1,x_2}(x_3) = 1 \tag{4.23}$$

for any $x_3 \in [0, 1]$.

3. *The axiom "free wills"* claims if the state and society do not motivate to observe the law ($x_1 = 0$), and also the state and society is estimated by the employee negatively as not motivating to observe the law ($x_2 = 0$), then any (!)aspiration (intension) of the employee responsible for ensuring anti-terrorist safety will lead to commission of a real act, at the same time such act can be both lawful, and illegal (x), that can be shown in the following look [43]:

$$Sub_{0,0}(x) = x \tag{4.24}$$

for any $x \in [0,\ 1]$.

It is necessary to explain that the real act based on intension of the employee can be any just because it and is a concept of free will of the classical understanding that is accepted as is and is not subject to additional proof at set $x_1 = 0$ and $x_2 = 0$.

For the solution of a research problem on decrease in risk of emergence and commission of terrorist attacks at airlines by development of a method of selection of shots, the most relevant is application of a so-called "formula of the person" V. A. Lefebvre which conceptual essence and the proof is given further:

Earlier specified operator $Sub_{x_1x_2}(x_3) = X_1(4)$ is representable as function in the following look:

$$X_1 = f(x_1x_2x_3) \tag{4.25}$$

Considering that $f(x_1x_2x_3)$ it is linear to each of the variables, the following theorem is right:

Theorem *Follows from axioms of "trustfulness", "not evil intention" and "free will"* [43]*:*

$$f(x_1x_2x_3) = x_1 + (1 - x_1 - x_2 + x_1x_2)x_3 \tag{4.26}$$

Proof

1. The linearity of function $f(x_1x_2x_3)$ to each of the variables, allows to state it by means of classical multilinear algebra in a trilinear look that will look so [43]:

$$f(x_1x_2x_3) = a_0 + a_1x_1 + a_2x_2 + a_3x_3 + a_4x_1x_2 + a_5x_1x_3 + a_6x_2x_3 + a_7x_1x_2x_3 \tag{4.27}$$

   where $a_i(i = 0, \ldots 7)$—real numbers.
2. The axiom "*free will*" sets function [43]:

$$f(0, 0, x_3) = x_3, \tag{4.28}$$

   from where $a_0 = 0$ и $a_3 = 1$.
3. The axiom "*not evil intentions*" sets function [43]:

$$f(1,\ x_2,\ x_3) = 1, \tag{4.29}$$

   from where $a_1 = 1$ and $a_5 = -1$, $a_2 + a_4 = 0$ and $a_6 + a_7 = 0$
4. The axiom of "*trustfulness*" sets function [43]:

$$f(0, 1, x_3) = 0, \tag{4.30}$$

   from where $a_2 = 0$ and $a_4 = 0$, $a_6 = -1$ and $a_7 = 1$

Considering the above, we will present a formula (4.25) in the following form:

$$X_1 = x_1 + (1 - x_1 - x_2 + x_1x_2)x_3 \tag{4.31}$$

Further, for convenience we will transform the received formula (4.31) to the following look:

$$X_1 = 1 - (1 - x_3 + x_2x_3) + x_1(1 - x_3 + x_2x_3) \tag{4.32}$$

also we will enter support function

$$F(v_1, v_2) = 1 - v_2 + v_1v_2, \tag{4.33}$$

where v_1 and v_2 are real numbers.

Considering the entered support function, the formula (4.32) can be present in following a look:

$$X_1 = F(x_1, F(x_2, x_3)) \tag{4.34}$$

Let's enter value X_2, which can be characterized as ability of the employee of service safety to estimate own readiness to observe the law and an order, we will call it own measure of readiness of the employee and we will bring in a type of the following formula:

$$X_2 = F(x_2x_3) \tag{4.35}$$

As well as it specified in paragraph 3.1, the existing methods of selection of the candidates who are directly involved in ensuring anti-terrorist safety of airline are insufficiently effective, have certain shortcomings. The methodical recommendations offered further are developed based on the mathematical models allowing making selection of candidates proceeding from personal qualities (a binary and intentional method).

4.5.1 Binary Method of Making Decision on the Validity

As a result of conducting testing according to the questionnaire of SMIL, the got "*crude*" points compared with T-points on concrete scales are applied to classical calculation of results and interpretation. The answer of the subject to a question truly or not leads a concrete statement to charge of "*crude*" points. Nevertheless, the overall picture of a psy-condition of the personality which is built after the complex analysis of all scales is rather bulky, contains the enormous volume of information on the subject, cannot estimate specifically (it is meant as obtaining the answer to a simple question—"*yes*" or "*no*", "*1*" and "*0*", without the need for additional interpretation of results by the psychologist-diagnostician) tendency to these or those criminal or destructive acts, and the option of norm is set by the author of the collection of SMIL in rather wide range—from 30 to 70 points that in the real research was already listed above among shortcomings of this system of testing.

The author offers the following technique of use of the developed binary method of selection of shots:

(a) for the real research it is offered to separate classical testing according to the questionnaire of SMIL from performance of tasks of assessment of the validity of the candidate directly to the work connected with anti-terrorist safety. For a specification of data on the candidate within the real research it is necessary to mark out the most important personal qualities by certain criteria, in this regard use of statistical scientific base of the additional scales capable to range concrete human qualities is productive, offering candidates filling of the test on ten additional scales (Appendix A). Similar approach will allow to reveal the candidate's indicators on each of the concretized scales. Data on norm of these additional scales are supposed to be used as the scientific statistical material having justification which will be used and processed otherwise, than it is supposed at standard testing of SMIL;
(b) for increase in originality and reliability of testing, decrease in an opportunity at the alleged candidate to find "*right*" answers, it is necessary not to specify numbers of questions connected with the classical questionnaire of SMIL in forms of testing. Thus, the candidate will not be able to compare influence of the answer to a possible outcome, even, possessing keys to the standard reference book of SMIL and additional scales whereas to learn the positive directed answers on all 566 questions will be a task almost impracticable;
(c) an author's technique of selection of shots on the basis of V. A. Lefebvre's theory and classical Boolean algebra, allow to reduce influence of a subjective factor in the person of the person holding testings, allowing to issue the final answer in a look "*is suitable*" or "*it is not suitable*", "*1*" or "*0*", respectively.

Let's consider the books elected by the author additional scales in Appendix A [29] for application of entrance data of scientific statistical norm which will be involved further within the real research.

We adapt axioms of "trustfulness", "free will", "not evil intentions" and "essence of a formula of the person" V. A. Lefebvre for application within the solution of a task of improvement of quality of selection of the employees involved in ensuring anti-terrorist safety of airline and we will correlate the corresponding $x_{1...3}$, X_1 real human qualities, a psy-condition of the examinee.

Let's describe variable theories of V. A. Lefebvre as follows:

1. x_1—*a measure of pressure of the state and society declining the employee to commission of a lawful act*;
2. x_2—*value judgment by the employee of pressure of the state and society declining it to commission of a lawful act*;
3. x_3—*internal aspiration of the employee (intension) to make a lawful act*;
4. X_1—*a measure of readiness of the employee to make a lawful (illegal) act*;

The concept of "a formula of the person" according to V. A. Lefebvre (4.32) should be applied within the real research in the form of the following expression:

$$X_1 = 1 - (1 - x_3 + x_2x_3) + x_1(1 - x_3 + x_2x_3) \tag{4.36}$$

For carrying out mathematical calculations and obtaining the end results in the form of classical Boolean algebra, it is necessary to substitute instead of the accepted symbols $x_{1...3}$, digital values "1" and "0", that as a result will lead to obtaining the end result—what decision will be made by the subject—to become the helper of terrorists and to conduct a terrorist attack as a negative pole—"0", or to support a positive pole—"1", in the form of the state and the law, honestly to fulfill the functions on defense of peace and about. Obtaining results of testing of the candidate when carrying out selection, will allow to make the final decision on its validity to work on ensuring anti-terrorist safety of airline.

Let's conduct a research of candidates on the developed author's method given in Appendix B to the book. The data received as a result of testing on additional scales of the SMIL test are considered as the entrance information which is subject to processing for a binary method of making decision on the professional validity of the candidate, and the theory V. A. Lefebvre is put in a basis of "making decision" on the validity of the specific candidate.

As an example of calculation for one of additional scales, we will believe that as a result of the choice of the statements relating to an additional scale crime, the examinee the subject—the man, gained 7 points. Let's substitute values of "crude" points, mean-square ($M = 11.28$), and mean-square deviations ($\sigma = 4.23$) in a formula and we will make calculations on a technique formula L. Sobchik [29]:

$$\gamma = 50 + \frac{10(7 - 11.28)}{4.23} = 40 \tag{4.37}$$

The received result 40 will be the value of T-points which is subject to assessment (further, entering or not entering rated range—"1" or "0", "Enters norm" or "Does not enter norm", respectively).

The analysis of an additional scale which is carried out in Appendix B "Crime" showed that the technically feasible interval lies in the range from 23 to 101 T-points. Proceeding from V. A. Lefebvre's theory, we will accept it which can be set by the well-known golden ratio of Fibonacci—number 1,618 as the most suitable value, having applied it to the smallest value of range—to number 23. We will accept the received number 37 as the most suitable result (to the best—minimum). Also we will allow the top step of hierarchy (to the head of airline, HR department and so forth) to set a tolerance—an error interval that, in the experimental purposes, proceeding from at most 101 T-points we will establish by value in the form of +10 and −10. Thus, the most suitable candidate has to gain from 27 to 47 points to become suitable ("1") on this scale, or not suitable ("0"),—if his results of T-points are less than 27 or higher than 47 points. Data on scales differ and are based on scientific statistical base of additional scales, at the same time processing of ten scales which are obviously possible for using without completions is given in Appendix B.

Table 4.1 Table of correspondences of measures to scales

№		Measure		Scale	
1	x_1	Pressure of the state and society, inducing the candidate to make a lawful act	L_1	"Social responsibility"	
2	x_2	Value judgment (candidate) of pressure of the state and society towards commission of a lawful act	L_2	h_1	"Crime"
				h_2	"The conscious expressed hostility"
				h_3	"Immorality"
				h_4	"Bias"
				h_5	"The double standard of the personality"
7	x_3	The internal aspiration (intension) of the candidate to make a lawful act	L_3	k_1	"Line of responsibility"
				k_2	"Relation to work"
				k_3	"Altruism"
				k_4	"Trustfulness"

Further, we will pass to need of compliance of results—theories of "a formula of the person" V. A. Lefebvre and additional scales of the questionnaire of SMIL for what the author's table of correspondences of measures to scales (Table 4.1) is used:

At correlation of the theory of V. A. Lefebvre to additional scales of the questionnaire of SMIL, reference by the principle x_1 to L_1 like as ("1 to 1"), and compound references which are relevant for x_2 to L_2 and x_3 to L_3. Correlation of elements, at which to one variable V. A. Lefebvre's systems there correspond several additional scales at once (see item 2–3 of Table 4.1), of the SMIL system, it is necessary to carry out on formulas:

$$L_2 = \frac{(h_1 + h_2 + h_3 + h_4 + h_5)}{z}. \tag{4.38}$$

where z is quantity of h of elements (in relation to—5).

Respectively for L_3 the formula takes the following form:

$$L_3 = \frac{(k_1 + k_2 + k_3 + k_4)}{e}. \tag{4.39}$$

where e is quantity of k of elements (in relation to—4).

For realization of an example of work of a binary method of selection of shots, the author invited for testing the acting employee of security service who is directly involved in fight against terrorism who anonymously took part in testing according to Appendix B in this connection the following results specified in the Table 4.2 were received:

Table 4.2 Results of testing of the employee on Appendix B

№		Measure			Scale		
1	x_1	Pressure of the state and society, inducing the candidate to make a lawful act	$L_1 =$	1	"Social responsibility"		
2	x_2	Value judgment (candidate) of pressure of the state and society towards commission of a lawful act	$L_2 =$	1	$h_1 =$	1	"Crime"
					$h_2 =$	1	"The conscious expressed hostility"
					$h_3 =$	1	"Immorality"
					$h_4 =$	1	"Bias"
					$h_5 =$	1	"The double standard of the personality"
3	x_3	The internal aspiration (intension) of the candidate to make a lawful act	$L_3 =$	1	$k_1 =$	0	"Line of responsibility"
					$k_2 =$	1	"Relation to work"
					$k_3 =$	1	"Altruism"
					$k_4 =$	1	"Trustfulness"

For conclusions about the examinee the subject (candidate) the mathematical apparatus is applicable. Let's present the finished formula of the subject to choose a positive or negative pole of the actions in the following look (4.32):

$$X_1 = 1 - (1 - x_3 + x_2 x_3) + x_1(1 - x_3 + x_2 x_3) \tag{4.40}$$

Let's substitute the received digital values in option of "a formula of the person" by V. A. Lefebvre:

$$X_1 = 1 - (1 - 1 + 1 * 1) + 1(1 - 1 + 1 * 1) = 1 \tag{4.41}$$

The received results allow to draw the following conclusion on the examinee the subject (interpretation of results of a research), to take out the corresponding conclusion on the candidate (sample):

CONCLUSION ON THE CANDIDATE (SAMPLE)

The man (*the age, education, and a position according to biographical particulars, is given further in a concrete case—anonymously).*

Are revealed: rated indicators of social responsibility, rated indicators of the generalized criterion (5 in 1) unreliability—crime, the conscious expressed hostility, immorality, bias, the double standard of the personality, rated indicators of lines of the generalized criterion of reliability (4 in 1)—the relations to work, altruism and trustfulness, at the same time is noted a negative indicator of the criterion of lines of responsibility.

Conclusion: ***the candidate is fit for work in the sphere of ensuring anti-terrorist safety of airline.***

To check for reliability and a possibility of application of the developed binary method of making decision on the validity of candidates, the author conducted a research on two groups (further—groups "A" and "B"), the acting employees:

(A) the employees who are directly fighting terrorism (from among the persons serving in the security service of the Russian Federation) with a length of service on a profile—not less than 5 years. Group as a part of 50 people;
(B) the employees having superficial skills and knowledge of the field of fight against terrorism (from among the persons who are engaged in operational work in the Internal Affairs Service of the Russian Federation) with a length of service on a profile—not less than 5 years. Group as a part of 50 people.

The principle of conducting testing on two brought groups is based that, as we know, the most serious of existing requirements at selection was imposed to persons from group "A" initially at the device on service, then to employees of group "B" (on whose level of training, in general, it is possible to judge some average general meaning of indicators characteristic of the serving usual divisions of the Department of Internal Affairs, Rosgvardiya, civil PSF and SAB and also the ordinary staff of the Russian Federal Border Service). Respectively, the hypothesis assumes if the developed methodical recommendations within a binary method are efficient, surveyed from group "B"—have to be objectively less adapted on personal by qualities for similar work, than employees from group "A" that it is possible to prove carrying out a research of groups and also a comparison of results between them.

For depersonalization and nondisclosure of data on duty stations and departmental accessories, observance of the current legislation, surveyed both groups did not specify the personal data, division was made only on groups, without a possibility of the establishment what questionnaire concerns the particular person. After conducting testing, the results of questioning were digitized and transferred to numerical values for a possibility of processing by a binary method according to the developed technique given to Appendices B.

The conducted research of group "A" allowed to draw a conclusion that from 50 of surveyed (Fig. 4.1):

- 46 employees—recognized as suitable;
- 4 employees—not suitable.

Fig. 4.1 Data on the validity of faces of group "A"

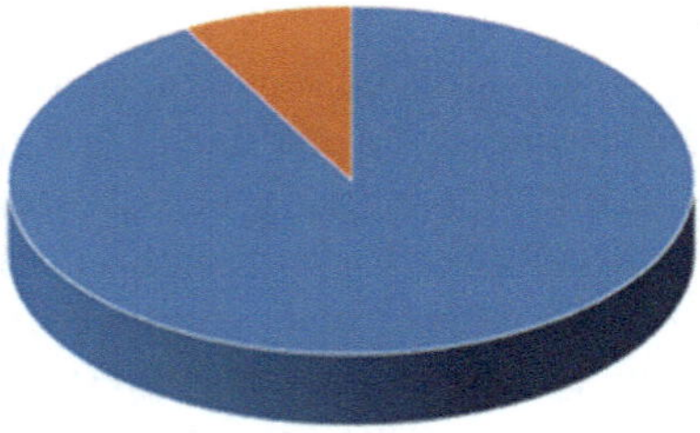

The received results of a research of group "A" allow to be convinced of a possibility of application of the developed method that is confirmed by an insignificant indicator (4 people), not of fit employees that objectively is true, considering that the examined persons could be in a bad psy-emotional condition or to be subject other, independent and sudden factor directly at the time of testing passing, testing errors are also not excluded.

The conducted research of group "B" allowed to draw a conclusion that from 50 of surveyed (Fig. 4.2):

- 11 employees—recognized as suitable;
- 39 employees—not suitable.

The received results of a research of group "B" also allow to draw a conclusion on reliability of the values received with application of the developed method. The employees having superficial skills and knowledge of the field of fight against terrorism (from among the persons who are engaged in operational work in the Internal Affairs Service of the Russian Federation) are objectively insufficiently trained for similar work (only 11 employees successfully coped with testing passing), in difference with employees from group "A" who are engaged in concrete activity on a regular basis, professionally.

Considering the conducted research of groups "A" and "B", their comparative results, it is possible to make the assumption that results of testing of employees of SAB of airline and various PSF can be much more the worst, than results of group "B", considering even more considerable difference in personal qualities of candidates. At the same time, it is necessary to consider that the developed method

Fig. 4.2 Data on the validity of faces of group "B"

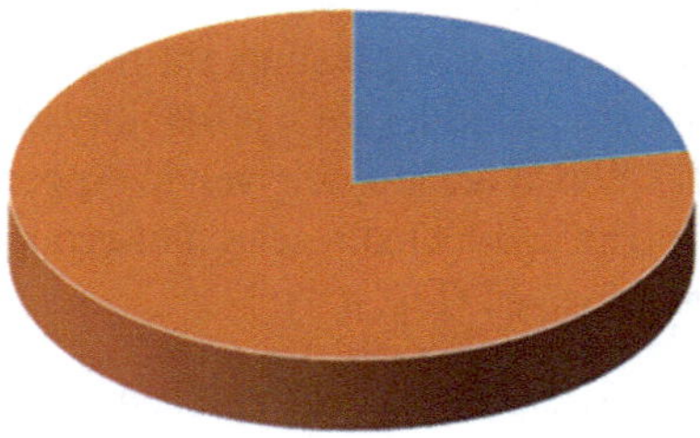

allows the top level of hierarchy to regulate the range of "norm" of candidates that can affect the results of selection considerably.

Results of the conducted research of groups confirm adequacy of the developed binary method of making decision on the validity of the employee involved in ensuring anti-terrorist safety of airline allow to draw an unambiguous conclusion on a possibility of the admission (not the admission) of the specific candidate to work on ensuring anti-terrorist safety. The received results also allow to draw a conclusion on the existence of enough serious personal requirements for candidates for the positions connected with ensuring anti-terrorist safety at which even not all acting staff of safety could pass successfully test, nevertheless—the vast majority on the basis of a new binary method coped with passing of testing successfully.

4.5.2 Intentional Method of Making Decision on the Validity

The aspiration of the person to trust—within the real research considered it "intension", that is internal orientation, tendency to a certain object, the purpose. In the general philosophical aspect, the concept "beliefs" is that truth existing irrespective of need of its proof and an explanation in logical ways. In other words, the belief is unconditional for the subject and is accepted by it unconditionally, as a rule, in a type of existence defined, created (a comment—or imposed by someone?) relations between the subject and subject of belief. The concept "beliefs" is also often used as religious belief in the understanding of the concrete doctrine (Christian, Muslim, Judaic belief). The religious belief, in the basis, is under construction on the belief of the subject in a certain reality and tangibility supernatural, as so predominating source of religious belief a certain deity, nonexistent (and never earlier not existing) the illusory "guru" capable through the "terrestrial intermediaries" acts (followers, preachers, etc.) to carry "intimate knowledge" of rules of life on the world and to send on "true" (in the understanding of such deity and also, perhaps, in the mercenary interests of followers and preachers) to a way. It should be noted that at the general understanding of religious belief as certain positive (temper) doctrine which the subject can choose "on himself" also religious radicalism and extremism which result is an education of various sects and cults of destructive sense meets. The purpose of development of similar communities is involvement of people to performance of certain actions for the benefit. there is nobody deities, counting on receiving for itself as an award—the extraterrestrial benefits and eternal life after physical death, an enlightenment and understanding of the nature of the universe and so forth, at the same time, for achievement of the goals can be required to make the acts claiming the lives of innocent people—for example—terrorist attacks, and, to sacrifice people of other belief. In the real work, the author does not suppose categorization, reference of this or that religion to more or less incline to such radical actions and also identification of a concept of religion and belief, and opens absolutely different, mathematical and philosophical, and knowledge-intensive view on this perspective. Investment of the employer with the ability to disclose the essence of intension of the specific candidate for the position connected with ensuring anti-terrorist safety at airline will allow to present the fullest picture about true aspirations, intentions of the similar person to commission of these or those actions.

Considering scope of the real research, we will take for the general concept of belief, in the calculations described further, belief of the candidate for a position directly to the state and the law as a positive pole and it is "good", to a negative pole—we will refer belief in terrorism and commission of "bad" acts.

But, before passing directly to the determination of tendency—"intension" to these or those actions at the candidate, it is necessary to find out who exactly of candidates will be able to make a choice who has such congenital or acquired lines of temperament which are applicable in a concrete profession. The temperament influences the bases of behavioral model of the personality, possible predetermination of actions of the employee at emergence of emergency situations, it is necessary to consider temperament—a basis of any intension of the employee. Let's assume that people with a certain type of temperament (and, especially, extremely expressed lines of temperament) do not approach similar professional activity at all that we will consider in detail further in the text.

The word temperament (latin of "*temperamentum*") literally in translation means—appropriate mix or harmony of components. In the staticized concrete understanding, the temperament is a characteristic of the personality, resistant line which sets the general vitality of behavior of the person, connected at the same time directly with features of higher nervous activity. Founders of studying of temperament, by right it is possible to consider doctors of the antique period—Hippocrates and Galen assuming that in a human body there are individual differences between the chemical systems of an organism, which regulate behavior of the person. Similar interpretation formed a basis for the theory about the existence of one of four elements, the so-called "vital juice" in a human body, as predetermines temperament of each of us [33].

Categorization of temperaments considered conditionally in the following representation:

- sanguine persons differ in cheerful nature and general mobility;
- the grief, melancholy, and sensation of fear is peculiar to melancholics;
- are allocated to phlegmatic persons with quiet temper and sluggishness;
- choleric persons, in turn, differ in irascibility, tendency to scandals and fieriness.

It should be noted that extreme, especially bright definitions of types of personality according to Hippocrates-Galen, as a rule, diagnose as various mental disorders and are specified in MKB as diagnoses: to the choleric person impulsive frustration of the personality (a code on MKB-10–F60.30) is peculiar, to the phlegmatic person—schizoid frustration of the personality (a code on MKB-10–F60.1), to the melancholiac—disturbing frustration of the personality (F60.6), and in the sanguine person it is inherent a hypomania (a code on MKB-10–F30.0). The pronounced temperament of the personality, in most cases, will be diagnosed as a mental disorder [44].

Conditional categorization of all candidates for a position on four types of temperaments (on close to them, the mixed types), promotes the fullest and convenient differentiation on validity degree to performance of the activity connected with ensuring anti-terrorist safety. When determining the validity of the employee to the activity connected with ensuring anti-terrorist safety it is necessary to mark out the main lines of the most important and demanded for professional activity of qualities, properties of the personality, for their correlation with a certain type of temperament and adoption of the further decision on an opportunity to be engaged in professional activity.

On the basis of carried out by the author researches, is possible to draw a conclusion on key personal qualities of the potential employee involved in ensuring anti-terrorist safety of airline. Such qualities are:

- speed of stay and acceptance of truth (from the point of view of morals, the law, a concrete situation, and so forth) decisions, the high speed of reaction at accidentally arising critical and force majeure;
- high intellectual and intellectual level (that can be in a certain measure to border on skill level, i.e. qualification quality), the developed informative qualities;
- resistance to stress, ability to control emotions and tranquility;
- high level of skill to communicate;
- ability to self-sacrifice for the good of the highest (in our case, it is undoubted—positive) the ideas and the purposes, understanding, and awareness of readiness to give the life for life of other people, for the good of public interests and values;

The ability to self-sacrifice for the good of the highest ideas and the purposes is rather a debatable subject, can be often unreasonably referred to radical fanaticism, but, nevertheless—depending on the orientation of these ideas and purposes—can also act for the good. In this research, the author considers ability to self-sacrifice as certain, the merit of the candidate for the corresponding position expressed in an opportunity to save at the price of the life—life of other people is unconditional. As a justification of the importance of similar quality for professional activity (without considering a question of negative or positive orientation of similar activity), it is possible to give an example of the Samurais of medieval Japan. Notes about the Samurai's lives under Tsunetomo Yamamoto's authorship stated in translation into book Russian "to Hagakure. Hidden in foliage. The code of honor of the Samurai", is described by selfless service to definite prime targets and also ability of the real Samurai on self-sacrifice [45]. In spite of the fact that self-sacrifice and readiness for death at the Japanese Samurais, in the extreme manifestation of personal honor, were expressed in commission of ritual seppuku, it is necessary to emphasize that the employee involved in ensuring the anti-terrorist safety of airline—always is at the last boundary of protection where the meeting with the suicide bomber can also become for it fatal. At the same time, also the fact that both warring parties—as the employee in the sphere of safety, so the terrorist—have to be equally psychologically ready is interesting "to go to death" for the achievement of the purposes and performance of objectives, as makes them similar, in the extreme manifestation, to an idealized image of the devoted and faithful Samurai for which ideal is the ability to leave the life resolutely, without thoughts, for the good of certain principles and the purposes.

Analyzing the above, it is necessary to draw a conclusion that the temperament sets a basis of actions of the person and predetermines his behavioral model. Undoubtedly, there can be an obvious question that the similar description of qualities of the personality is too idealized and, in practice, it is impossible to combine them in one person, and to find the suitable candidate, in that case, will become a task if not impracticable, then—extremely difficult. Nevertheless, knowledge-intensive approach to the solution of a difficult task by means of an idealization method (a comment to psychology) allows to find its solution. Approach to which the focus on the most suitable candidates and also their rather "tough" selection and categorization is peculiar, at all the possible complexity of carrying out,—according to the author will be able to lead to the best solution of an objective.

As the main mechanism of determination of temperament, the technique of the questionnaire of EPI of the famous German-British psychologist Hans Juergen Eisenk is capable to define temperament type by a two-factor model of an extroversion and a neuroticism [46] most successfully proved. The technique was developed by the scientist on the basis of the analysis of materials of inspection of 700 soldiers having neurotic diseases on which the scientific and statistical conclusions put in base of conducting testing were drawn. Testing for definition like temperament is held by means of the questionnaire during the work with which, the examinee needs to answer 57 questions—"yes" or "no". At interpretation of results of the questionnaire, the key to the questionnaire which correlates the answer to each specific question of the examinee to his tendency to an extraversion or a neuroticism is used, and also the indicator of sincerity of answers is considered, score is counted. The general analysis and interpretation of results is made by means of "Eisenk's circle" with imposing vertical and horizontal lines which crossing will show a point—the place of the examinee in the two-factor model.

It should be noted that according to G. Yu. Ayzenk, extreme degrees of temperament do not exist (besides, as well as it was stated above—similar can be the pathological state requiring special attention). Between four types of temperament about 16 intermediate forms as variations of the main types settle down [46].

On the basis of carried out researches in paragraph 3.1, it is necessary to draw a conclusion that unacceptable qualities of the candidate for a position of the employee involved in safety are in a projection to "Ayzenk's circle": easily upset, rigidity, inclined to reasonings, pessimistic, uneasy, aggressive, excitable, changeable, impulsive, carefree, easy. More, the specified negative lines belong to temperament of the melancholiac and choleric person. For making decision on impossibility of the admission of the specific subject to further psychophysiological tests formulation of the question about existence of extreme, especially bright states at which identification the candidate, anyway, will be defined as the person having a disease is necessary and it is not allowed to similar work. Obtaining results of a research of the candidate in which analysis, the mixed temperament type, decides on lines: restraint, discretion, judiciousness, goodwill, reliability, tranquility, sociability, responsiveness, initiative, activity that testifies about reference of his temperament to type of

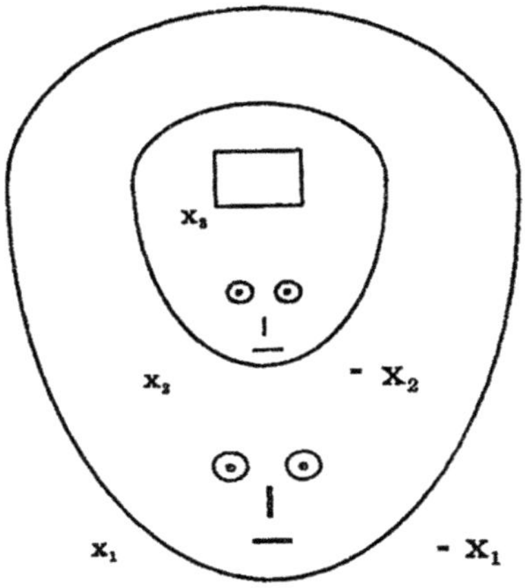

Fig. 4.3 Graphic metaphor of the theory of V. A. Lefebvre [43]

the phlegmatic person and sanguine person to a large extent, involves the admission to further psychophysiological tests, to the following step of the procedure of selection of shots.

Let's believe that previously selected as temperament with the prevailing above-stated lines (peculiar to sanguine persons and phlegmatic persons) candidates possess own aspiration to any action (further—an intentionality). According to V. A. Lefebvre's theory, there are two types of intentionality:

1. At an intentionality of the first type, the candidate for a position has a model of an external object, and this model performs function of an image of the world in an image of [43].
2. At an intentionality of the second type, the candidate for a position has a model of, and this model performs function of an image of in an image of [43].

The graphic metaphor of the theory is shown by V. A. Lefebvre in the following look (Fig. 4.3):

In spite of the fact that the general explanations were presented to variables in paragraph 3.2, is necessary slightly to change formulations and to state them in the following look as it is presented in V. A. Lefebvre's works [43]:

x_1—*sensory perception by the candidate of the world;*

x_2—*the generalized image of the world;*

x_3—*model of an object to which action of the candidate is directed;*

X_2—*an image of the candidate for;*

X_1—*the candidate (as he is, in reality).*

It is obvious that under these variables the real sensual experiences of the candidate connected with his views of the world, the state, the law and an order, of the relation to work and relatives and also hundreds more of individual parameters which can differ in a considerable image for each of us disappear. Besides, experiences can have a certain force, to be both superficial, and deep, depending on the general tension of the candidate, his psychophysiological state, and, undoubtedly, conditions of real surrounding reality. There are no doubts that when finding the candidate in the bad world which is represented to that for him and also at low opinion the candidate about himself, at it will have into the account of weak experiences. Specified and also graphical representation in the form of a metaphor is shown us that not only the candidate, but also his image of can possess experiences. Are also peculiar to an image of experience of both the world, and themselves. Thus there is a subtask on introduction to a conceptual framework of concepts of deep and superficial experience which can be applied as the candidate to themselves and the world and also can arises also at a sample of in the candidate that can be shown the following expressions [43]:

$$d = 1 - x_2 - deep\ experience\ by\ the\ candidate\ of\ the\ world; \tag{4.42}$$

$$D = 1 - X_2 - deep\ experience\ by\ the\ candidate\ of; \tag{4.43}$$

$$s = 1 - x_3 - superficial\ experience.\ world\ in\ the\ way\ of\ the\ candidate; \tag{4.44}$$

$$S = 1 - x_3 - superficial\ experiences\ in\ the\ way\ of\ the\ candidate; \tag{4.45}$$

Using the stated data, we can turn to a question of belief as to a basis of an intentional method of making decision on the validity of the candidate involved in ensuring anti-terrorist safety of airline. The belief is the subject's relation to an object, sometimes and an object to the subject (that, perhaps, meets much less often). The mathematical essence of similar can be express simple the relations: Trusts M in N. At the same time, it is logical to assume that the power of faith can be characterized criterion of depth (force)—intensity, the intensity will show as far as M N is ready for the sake of the belief—to observe the law, to offer itself and also to make still is multiple acts which depend only on intensity of such belief. Thus, it is necessary to set variables for designation of belief. Let's call a variable r—*the index of belief for an intentionality of the first type and,* respectively R—*the index of belief for an intentionality of the second type.*

According to the intentionality types described above, the first type of intensity of belief will look as follows [43]:

$$r = \frac{d}{s}, \tag{4.46}$$

and for an intentionality of the second type:

$$R = \frac{D}{S}, \tag{4.47}$$

Further, we will show the typical set expressions characterizing intensity of belief, for example:

$$r < 1 < (R < 1) - \textit{the belief is rather weak, than strong;} \tag{4.48}$$

$$r > 1 > (R > 1) - \textit{the belief is rather strong, than weak;} \tag{4.49}$$

$$r = 1 - \textit{a balance point between strong and weak belief;} \tag{4.50}$$

$$R = 1 - \textit{a balance point between strong and weak belief;} \tag{4.51}$$

It should be noted that hardly in the real world, in practice, it is possible to apply option with equilibrium value between weak and strong belief at which the index of belief is equal to *1*. It is possible to apply similar value to modeling of a certain conditional situation with the candidate at which it is necessary to show neutrality (balance) of his belief, nevertheless—similar theoretical modeling will have nothing in common with the real situation at the real personnel selection as any person is in reality under constant pressure from the state and society and also separate unreliable elements, under the general action of what, the index of belief inevitably begins to aspire to this or that pole.

Let's assume that the candidate has to note a point on a scale, at the same time one end of a scale has to be correlated to value *1*, another to value *0*, respectively (Fig. 4.4).

It is possible to accept the fact that information at the level of feelings on the subject of how people around really will begin to treat it the candidate has no that in that case can be shown expression $x_1 = \frac{1}{2}$.

To the candidate there will correspond in that case expression: $x = \frac{1^{x_2^x}}{2}$ (4.52)

Considering that at such task the candidate focused generally on the external relation to it, using the concepts of the index of belief and a formula given in paragraph 3.2 it is possible to receive the following expression [43]:

$$r = \frac{d}{s} = \frac{1 - x_2}{1 - x}. \tag{4.53}$$

We will take for "a belief pole" that which corresponds to number *1*. At the same time X it is possible to express through r, as it was already mentioned in Lefebvre's

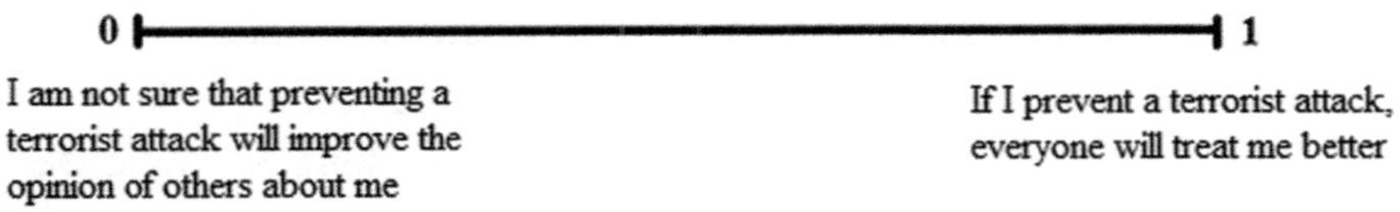

Fig. 4.4 Scale in intention

works, respectively from formulas (4.52, 4.53) we will express [43]:

$$x = \frac{r - 2 + \sqrt{4 + r^2}}{2r}. \tag{4.54}$$

Expression (4.54) also objectively to show that the belief of the candidate is stronger—the higher it ability to choose a positive pole. Than weaker it belief—that high probability that both the positive, and negative pole can choose the candidate that can be represented in the following look [43]:

If $r \to \infty$, then $x \to 1$ monotonously;
If $r \to 0$, then $x \to \frac{1}{2}$ monotonously.

It is also interesting to pay attention, at value $r = 1$ (as it was noted above, it is rather an a value for theoretical model), readiness of the candidate to choose a positive pole is equal to the well-known golden ratio that is confirmed by the following calculation:

$$x = \frac{\sqrt{5} - 1}{2} = 0.618. \tag{4.55}$$

If in an example above, the candidate focused on the outside world, then we will review below an example when he is puzzled with the solution of an internal question. The candidate has to choose a point on a scale, at the same time one end of a scale has to be correlated to value *1*, another to value *0*, respectively (Fig. 4.5).

Let's assume that the candidate does not possess information on as far as he is capable of violation of the law that can be expressed, respectively, in a look $x_1 = \frac{1}{2}$.

Considering that it is focused on the solution of an internal question, respectively—on himself, using the concepts of the index of belief and a formula given in paragraph 3.2 it is possible to receive the following expression:

$$R = \frac{D}{S} = \frac{1 - x_2^x}{1 - x} = \frac{2x - 1}{1 - x}. \tag{4.56}$$

Let's carry out transformation and we will present x through R in the following form:

$$x = \frac{R + 1}{R + 2}. \tag{4.57}$$

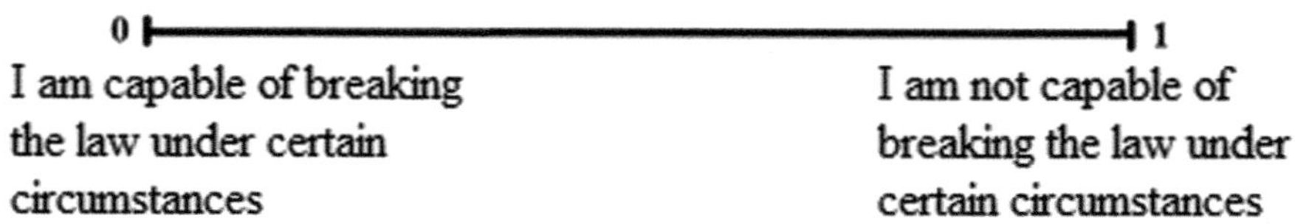

Fig. 4.5 Scale of intention

Let's receive the following results:

If $R \to \infty$, then $x \to 1$ monotonously;
If $R \to 0$, then $x \to \frac{1}{2}$ monotonously.

The logicality and validity of expressions 4.56 and 4.57 also do not raise doubts and is identical on the sense to the calculations which are carried out above (4.53, 4.54). As well as in expressions (4.53, 4.54), it is necessary to pay attention that at $r = 1$, the condition of the subject also corresponds to golden ratio, and at $R = 1$ size $\frac{2}{3}$ [43].

Thus, the author to application offers the following algorithm of carrying out an intentional method of selection of shots:

(a) it is required to make preliminary selection at the heart of which need of identification of temperament of the candidate is put. Candidates with the mixed temperament type, with the prevailing lines are the most suitable for this work (service) as it was stated above: restraint, discretion, judiciousness, goodwill, reliability, tranquility, sociability, responsiveness, initiative, activity that testifies about correlation of the candidate of temperament to type of the phlegmatic person and sanguine person to a large extent. In the real research it is not given a software testing technique to the personal questionnaire of EPI of G. Yu. Ayzenk in connection with its broad availability in profile literature. The candidates suitable as temperament are allowed to further tests;
(b) candidates are offered to answer a number of questions, at the same time questions can be divided conditionally into two types (that, eventually, matters only for the organizer of tests):

    (1) experiences of the candidate about the outside world (d);
    (2) experiences of the candidate about inner world (D).

Author's intentional technique of selection with questions books are given in Appendix B to this.

It should be noted that the answer to a question is presented by the candidate in the point form on a scale with set by a step, at the same time the scale has two poles: at the left negative—"0", on the right positive—"1", respectively. For obtaining concrete numerical data—the candidate's indicators, on each question the scale with the corresponding numbers and an interval is entered (from 0 to 1, with a step 0,1). As results the corresponding numerical values—d for external and D—for internal experiences, respectively, will be determined by each of the asked questions. The formulation of questions is constructed so that the candidate had as little as possible aspiration to specify extreme values. In case of obtaining exclusively categorical (0 or 1) answers on suppressing the majority of questions, without intermediate values, the test result has to raise doubts in reliability at checking, at the same time similar can demonstrate as attempt of the candidate to obviously distort results, and about his carelessness to the procedure of passing of testing. In general, similar approach when testing will allow to transform sensual intensions of the candidate for conditional numerical values which the candidate estimates and displays on a scale that is necessary for further assessment.

(c) the top step of hierarchy (the employer, the employee of HR department, the psychologist) sets an admissible interval of intension which demonstrates to positive intensions of the candidate for his professional activity that, eventually allows to make the decision on a possibility of employment (service)—"is suitable" or "is not suitable", in case of existence of other personal tests—to resolve a question of the admission of the candidate to the following level of testing. As an example, considering the theory of V. A. Lefebvre, rather existing range from 0 to 1, the author made the decision to use the golden ratio of Fibonacci equal 0.618. In the given example the candidate will be recognized not suitable for the corresponding work if his intension ≤ 0.617 whereas at values of intension ≥ 0.618—the candidate is fit.

For realization of an example of work of an intentional method of selection of shots, the author invited for testing the acting employee of security service who is directly involved in fight against terrorism who anonymously took part in testing by means of Appendix B in this connection the following results of testing were received (Table 4.3).

The final result of selection by an intentional method of selection of shots: average total value of indicators on scales (d, D): 0.79.

Conclusion: the candidate is fit.

Table 4.3 Results of testing

Calculation of results on a scale (experiences of the candidate about the outside world = d)		Calculation of results on a scale (experiences of the candidate about inner world = D)	
№ of answer	Result (digital value)	№ of answer	Result (digital value)
1	0.9	1	0.9
2	0.3	2	0.50
3	0.7	3	1
4	1	4	0.9
5	1	5	0.9
6	0.9	6	0.5
7	1	7	0.8
8	1	8	0.5
9	0.6	9	0.7
10	0.9	10	0.8
Average result	0.830	Average result	0.660
Summing up (experiences of the candidate about the outside world = d)		Summing up (experiences of the candidate about inner world = D)	
Orientation of intention	Positive	Orientation of intention	Positive

Explanation: considering V. A. Lefebvre's theory, we will accept experiences of the candidate about the outside world (d) as earlier specified variable x_1—the measure of pressure of the state and society inducing the candidate to make a lawful act, at the same time experiences of the candidate about inner world (D) as its real intension to make an act, earlier specified variable x_3—a measure of internal aspiration (intension) of the candidate to make a lawful act. Similar allows to give both values of variables x_1 and to x_3 to "decisive" ("*1*"—a positive pole, "*0*"—a negative pole) according to the theory, i.e. to counterbalance their importance for decision-making and to give an opportunity of obtaining the general average result on scales by which and to make the final decision on the validity. At the same time, on scales of d and D, respectively, it is possible to apply numerical values of answers to questions to detailed studying of the identity of the candidate, outside automatic processing of results on a formula.

For check of reliability and a possibility of application of the developed intentional method of making decision on the validity of candidates, the author of the book conducted a research of two groups (further—groups "B" and "G"):

(B) the staff of Internal Affairs Service having superficial skills and knowledge of the field of fight against terrorism (from among the persons who are engaged in operational work in Internal Affairs Service of the Russian Federation) with a length of service on a profile—not less than 5 years. Group as a part of 50 people.

(G) persons, from among personnel of airline, not connected with safety (the technical and support personnel of airlines which is not involved in work of SAB), from among civilians. Group as a part of 50 people.

The principle of conducting testing on two brought groups is based that, as we know, to persons from group "B" initially at the device on service were shown to requirements, connected with need to provide legality and an order, to have certain personal qualities (justice, firm intentions, moral stability and so forth). Whereas the personnel from group "G"—are objectively not adapted for the work connected with safety, especially anti-terrorist, demanding certain personal qualities. Respectively, the hypothesis assumes if the developed methodical recommendations within an intentional method are efficient, surveyed from group "G"—have to be in the majority not adapted for the work connected with safety in difference from persons from group "B" that it is possible to prove carrying out a research of groups and comparison of results between them.

For depersonalization and nondisclosure of data on duty stations of staff of Internal Affairs Service and these workers from among personnel of airline, for observance of the current legislation about personal data, surveyed both groups did not specify the personal data, division were it is made only on groups, without a possibility of establishment what questionnaire concerns the particular person. After conducting testing results of questioning were digitized and transferred to numerical values for a possibility of processing by an intentional method according to the developed technique given to Appendices B.

The conducted research of group "B" allowed to draw a conclusion that from 50 of surveyed (Fig. 4.6):

Fig. 4.6 Data on the validity of faces of group "B"

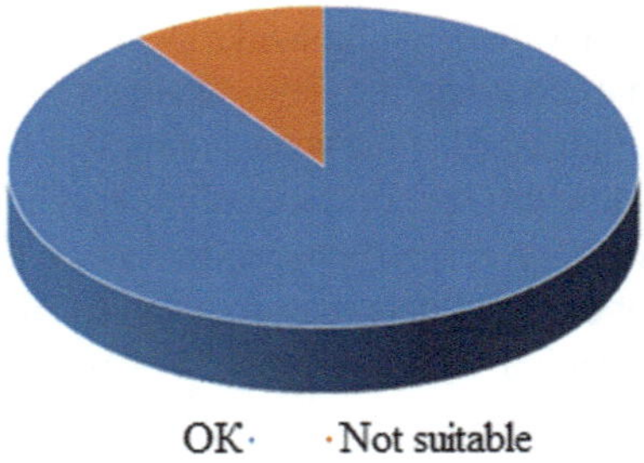

- 45 employees—recognized as suitable;
- 5 employees—not suitable.

The received results of a research of group "B" allow to be convinced of a possibility of application of the developed intentional method that is confirmed by an insignificant indicator (5 people) not of fit employees, it can be caused both by a certain error when conducting testing, and the actual discrepancy of specific employees to qualifying standards, professional burning out, etc.

The conducted research of group "G" allowed to draw a conclusion that from 50 of surveyed (Fig. 4.7):

- 7 workers—recognized as suitable;
- 43 workers—not suitable.

The received these researches of group "G" allow to be convinced applicability of the developed method. The group consisting of personnel of the airline, by the nature of activity which is not connected with safety (the technical and support personnel of airlines which is not involved in work of SAB) objectively is not adapted for the work connected with ensuring anti-terrorist safety does not possess the relevant personal requirements that led to obtaining results about worthlessness to such work of 43 of 50 people. Data on the validity of small subgroup of workers in number of 7 people, undoubtedly confirm their positive personal qualities at which they could apply for more important and responsible work.

The given results of a research of groups allow drawing a conclusion on a possibility of application of the above technique for improvement of quality of carrying

Fig. 4.7 Data on the validity of faces of group "G"

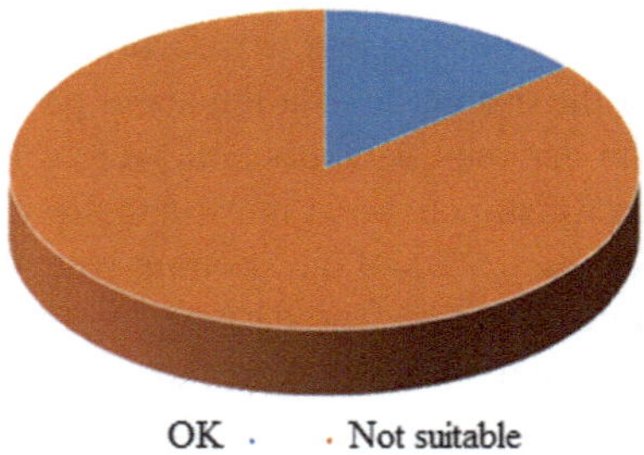

out selection of the employees involved in ensuring anti-terrorist safety of airline. The technique showed that a basis for the employee make this or that decision its intension could be, as predetermines commission of a certain act by it, proceeding from the conditional index of belief. The similar method can be to use as a part of complex psychological diagnostics, among other methods of selection of shots. Receiving at a stage of selection of shots by the representative of the employer of objective information from the candidate on his aspirations (intensions), considering reference of his personality to this or that type of temperament, will allow to warn possible negative acts which such employee can make, being employed (service).

4.6 Conclusion of the Chapter

1. Methods of selection of experts divided on personal and qualification. At the same time, the most important follow to consider personal qualities of the candidate, as without their appropriate level of development and strengthening of necessary qualification qualities is impossible.
2. The major factors exerting impact on the personality, behavior, and acts of the expert involved in ensuring anti-terrorist safety are:

   – *the environment factor* including impact on the identity of society, the immediate environment, and family;
   – *the factor of climate and food* defining impact on the identity of climatic conditions (nature) of the residence or work and connected from these inconveniences (or opposite);
   – *genetics*—as a biological factor, including process of the early birth and development of the child at even irresponsible age;
   – *a case*—the most difficult predicted factor depending on casual combination of circumstances.

3. The methods of a research of the identity of experts which are widely applied now are constructed, as a rule, on interpretation of results by specialists psychologists in a rather wide range, at the same time making decision on the validity of the candidate to a concrete kind of activity, is assigned to a subjective discretion of the only expert conducting a research. Similar approach inevitably leads as to reception at a position of the candidates not capable to carry out the functions, and to refusal in reception of persons, suitable for service, because of adoption of the wrong decision by the employer's expert when carrying out selection.
4. The developed probabilistic method of making decision on the validity of the expert shows the influence of the number of the checked qualifica[illegible]n signs on probability of the correct answers for candidates for positions of [illegible] involved in ensuring anti-terrorist safety with various qualificati[illegible]
5. The developed methodical recommendations within a binary m[illegible] decision on the professional validity of the experts involved [illegible] terrorist safety of airline allow to make the decision on expedien[illegible]

(service) of the employee, the most suitable for the specified functional activity, by delivery on the end of testing of definite answer—"is suitable" or "is not suitable".

The practical importance of methodical recommendations consists of small probability of the psychologist to incorrectly interpret results of a research of the candidate. This results from the fact that the known "formula of the person" V. A. Lefebvre allows to check results by means of classical axioms of psychology that, finally, will always lead to delivery of definite answer about the validity in the form of obtaining 1 or 0, is suitable values or "is not suitable", respectively, at the same time, the range of "norm" is regulated by the top step of hierarchy concerning requirements of airline that in earlier applied methods of testing was not used.

6. Takes the submitted methodical recommendations within an intentional method of selection of the experts connected with ensuring anti-terrorist safety of the air enterprise for a concept of belief, belief of the subject directly to the state and the law as a positive pole and *it is "good"*, and to a negative pole—belief in terrorism and commission *of "bad"* acts. Along with determination of temperament of the candidate, at similar approach and calculations, it is shown that a basis of the employee to make this or that decision it will be exclusive its intension—aspiration which predetermines the choice. Preliminary identification of objectively negative intensions of the candidate will allow predicting and preventing negative consequences.

References

1. Vorobyov VV (2012) Problems of prevention of collision of civil aircrafts in the operated flight. In: VV, Mozolyako EVM (eds) Sparrows: scientific bulletin of the Moscow state technical university of civil aviation, (183), pp 109–113
2. The order of the Ministry of Transport of the Russian Federation of 21.08.2014 No. 231 "About the approval of Requirements to knowledge, abilities, skills of forces of ensuring transport safety, to personal (psychophysiological) qualities, level of physical training of separate categories of forces of ensuring transport safety, including features of check of compliance of knowledge, abilities, skills of forces of ensuring transport safety, personal (psychophysiological) qualities, level of physical training of separate categories of forces of ensuring transport safety in relation to separate means of transport" (It is registered in the Ministry of Justice of the Russian Federation 09.10.2014 N 34278)//Reference and information system Consultant+
3. Borzova AS (2018) Training Problems in civil aviation. The scientific readings on aircraft devoted to N.E. memory of Zhukovsky, No. 6, pp 29–43
4. Guzy AG (2018) Methods of a posteriori of the aviation incidents caused by "A human factor". In: Guzy AG, Lukshin AM, Mayorova YuA (eds) All-Russian institute of scientific and technical information of RAS, No. 1, pp 7–18
5. Zubkov BV (2007) Aviation security (in 3 parts). In: Zubkov BV, Sakach RV, VA (eds) M's Skeletons: MSTU of HECTARE, pp 84, 76, 88
6. aex.ru [URL]: Balyasnikov VV. An organizational factor in a control system of safety of flights. http://www.aex.ru/docs/4/2014/8/7/2091/. Accessed 01 Apr 2019

7. dspk.cs.gkovd.ru [URL]: Doc 8973/9 RESTRICTED. Guide to aviation safety. It is approved as the Secretary general and published from his sanction. International organizations of civil aviation (ICAO). The edition ninth—2014. http://dspk.cs.gkovd.ru/library/data/8973_cons_ru_ruk__vo_po_ab_izd_9_e_2014g_.pdf. Accessed 12 Feb 2020
8. Federal law "About Transport Safety" of 09.02.2007 No. 16-FZ//Reference and information system Consultant+
9. "The federal system of ensuring aviation safety (The national program of aviation safety)" (it is approved by the Interdepartmental commission on aviation safety, safety of flights of civil aviation and simplification of formalities 15.10.2019)//Reference and information system Consultant+
10. "The federal system of training of personnel in the field of aviation safety (The national training program in the field of aviation safety)" (it is approved by the Interdepartmental commission on aviation safety, safety of flights of civil aviation and simplification of formalities 04.04.2019))//Reference and information system Consultant+
11. The draft of the Order of Ministry of Labour and Social Protection of the Russian Federation "About the approval of the professional standard "The Expert of Transport and Aviation Safety on Air Transport of Civil Aviation" (as of 06.02.2020) (ID of the project 01/02/02–20/00099272 is prepared by Ministry of Labour and Social Protection of the Russian Federation)//Reference and information system Consultant+
12. The resolution of the Government of the Russian Federation of 06.12.2012 No. 1259 "About the approval of Rules of professional psychological selection on service in Internal Affairs Service of the Russian Federation"//Reference and information system Consultant+
13. Rosgvardiya's order of 16.11.2016 N 357 "About the approval of the Instruction about the organization and carrying out professional psychological selection in troops of national guard of the Russian Federation" (It is registered in the Ministry of Justice of the Russian Federation 12.12.2016 N 44651)//Reference and information system Consultant+
14. The order of FSB of Russia of 13.04.2011 No. 151 "About the approval of the Instruction on the organization and carrying out professional psychological selection in staff of Federal Security Service" (It is registered in the Ministry of Justice of the Russian Federation 02.06.2011 N 20926)//Reference and information system Consultant+
15. Zanina TM (2009) About problems of psychological aspect of staff recruitment of Internal Affairs Service. In: Zanina TM, Razuvaev NI (eds) Obshchestvo and right, No. 4(26), pp 286–289
16. Strigunenko YuV (2007) To a question of some objective and subjective factors of professional deformation of the identity of the police officer in Russia. Society: policy, economy, right, No. 2
17. Myagkikh NI (2004) Professional psychological selection in Internal Affairs Service of the Russian Federation: concept, principles, criteria, technologies: yew. edging. psy. sciences: 19.00.06: 12.2004 are protected: utv. 11.2004.—SPb, 174 p
18. Razhivina IV (2013) Problematic issues of formation of personnel structure of territorial authorities of internal affairs in modern conditions. Jurisprudence and law-enforcement practice, No. 2(24), pp 96–100
19. Rydchenko KD (2016) The prospects of improvement of professional selection on service in Internal Affairs Service. In: Rydchenko KD, Yurist PRA (eds) Pravoved, No. 3(76), pp 115–120
20. Russell D (2019) Pyramid of requirements across Maslou. In: Russell D, Kohn R (eds) LLC Kniga po Trebovaniyu elektr. edition, 40 p
21. The charter (Constitution) of the World Health Organization (WHO) (it is accepted in New York 22.07.1946)//Reference and information system Consultant+
22. Hachikyan PP (2016) About the factors influencing the level of anti-terrorist training of staff of service of aviation safety//Natural and technical science, No. 4(94), pp 116–125
23. Hachikyan PP (2016) The all-military causes of corruption in the Armed Forces of the Russian Federation//the Young scientist, No. 11. pp 1371–1374
24. Hachikyan PP (2016) Corruption crimes in the Armed Forces of the Russian Federation//Questions of the Russian and international law. No. 7, pp 87–95

25. Federal law "About Anti-corruption" of 25.12.2008 No. 273-FZ//Reference and information system Consultant+
26. Hachikyan PP (2016) Organization of specialized anti-terrorist group in staff of the existing service of aviation safety at airline//Works of the international symposium "Reliability and quality" No. 1, pp 193–194
27. Hachikyan PP (2017) Anti-corruption in the Armed Forces of the Russian Federation. Monograph. Pero Publishing House, 109 p
28. Zaytseva EN (2001) Guide to psychological ensuring selection, preparation and professional activity of flight and dispatching structure of civil aviation of the Russian Federation. Parts 1–5. Air Transport Publishing House, 280 p
29. Sobchik LN (2007) The Standardized multiple-factor method of a research of the identity of SMIL (MMPI) Practical guidance. Speech, 224 p
30. Berezin FB (1976) Metodika of a multilateral research of the personality (in clinical medicine and psychohygiene). In: Berezin FB, Miroshnikov LS, Rozhanets RV (eds) Meditsina Publishing House, 176 p
31. Zaytsev VP (1981) Variant of psychological test of Mini-Mult.—Psikhol. zh., No. 3, pp 118–123
32. Kincannon I (1968) Prediction of standard MMPI scale scores from 71 items: The Mini-Mult. J Consult Clin Psychol 32:319–325
33. Ayzenk GYu (2016) Ayzenk's tests. IQ. Megamind. Open possibilities of the intelligence. (Shtayerman V. Lane under the editorship of Oleynik E). Eksmo-Press, 256 p
34. Solovyov S (2016) Personnel technologies. (electron. book). Multimedia Publishing House of Strelbitsky
35. Presidium of the Supreme Court of the Russian Federation. The review of cassation practice of judicial board on criminal cases of Russian Armed Forces for the 2nd half-year 2012 (point 5.2.1).//Reference and information system Consultant+
36. Ushakov AYu (2015) International experience of application of a polygraph in fight against crime. Problems of law-enforcement activity, (4), pp 110–112
37. Hell L (2006) Theories of the personality. Hell L, Lane ZD, Melenevskaya S, Viktorov D (eds) St. Petersburg, 217 p
38. Novikov DAVA (2016) Lefebvre and game theory: calls for a mainstream. Reflexive processes and management, No. 1–2, pp 54–58
39. habr.com [URL]: Use of statistical criteria at the solution of problems of detection in radio engineering. https://habr.com/ru/post/301476/. Accessed 07 Oct 2019
40. Hachikyan PP (2016) Criteria for evaluation of wrong actions of aviation personnel//Current state of natural and technical science, No. XXII, pp 37–41
41. othereal.ru [URL]: A hypothesis of the German scientists—actions are programmed. https://othereal.ru/gipoteza-nemeckix-uchenyx-dejstviya-zaprogrammirovany/. Accessed 05 Aug 2019
42. Lefebvre VA (1968) Algebra of the conflict. In: Lefebvre VA, GL (eds) Residents of Smolensk—1st prod
43. Lefebvre VA (1991) Formula of the person: contours of fundamental psychology. The lane with English. 108 p
44. International classification of diseases (MKB-10) (utv. The order of the Russian Ministry of Health of 27.05.97 N 170) (with the last changes and additions)//Reference and information system Consultant+
45. Tsunetomo Y (2019) Hagakure. Hidden in foliage. Code of honor of the Samurai. (In: Mizinin T, trans). Centrpoligraf Publishing House, 319 p
46. Raygorodsky DYa (editor originator) (1998) Practical psy-diagnostics. Techniques and tests. Manual. BAHRAH publishing House, Samara, 672 p
47. Elisov LN (2006) Quality of vocational training of aviation personnel and safety of air transport. Monograph. ITsPKPS Publishing House, 244 p

Conclusion

In the present book, the models developed by the author, methods, and methodical recommendations allowing estimating and reducing the risk of realization of terrorist threats in airline are in detail described. The result of a research will be achieved by the improvement of the organization of productions and optimization of the distribution of a financial resource at the realization of system measures in relation to separate organizational measures of system of anti-terrorist safety of the air enterprise.

In the course of research work, the author received the following main results:

- Models of distribution of the set reliability between elements of a highly reliable system of ensuring anti-terrorist safety by the criterion of use of the minimum financial resource.
- Models of distribution of a limited financial resource between elements of a highly reliable system of ensuring anti-terrorist safety by the criterion of ensuring the maximum probability of protection.
- Model of assessment of probability of realization of the act of terrorism depending on quantity and reliability of boundaries of protection.
- Model of "system of ensuring anti-terrorist safety of airline".
- A method of selection of the experts involved in ensuring anti-terrorist safety of airline for qualification qualities.
- Methodical recommendations about the selection of the experts involved in ensuring anti-terrorist safety of airline for personal qualities.

The received results allow:

- to provide the set reliability of the system of anti-terrorist safety of airline at the minimum financial resource;
- to provide the maximum reliability of the system of anti-terrorist safety of airline at the set financial resource (minimization of a financial resource and maximizing reliability);
- to estimate the probability of realization of the act of terrorism in airline depending on quantity and reliability of boundaries of protection;

P. P. Hachikyan, *Countering Air Terrorism*, Springer Aerospace Technology,
https://doi.org/10.1007/978-981-19-6172-4

- to use model of "the system of ensuring anti-terrorist safety of airline" at adoption of administrative decisions in the field of safety;
- to increase the quality of selection of the experts involved in ensuring anti-terrorist safety of airline in qualification qualities;
- to increase the quality of selection of the experts involved in ensuring anti-terrorist safety of airline in personal qualities.

Results allow to reduce risk of realization of terrorist threats in airline, emergence of the emergency situations, and adverse effects connected with it, to keep human lives and material values, and to provide stability to a social and economic and political situation in the world that testifies to the solution of a scientific task and achievement of the goal as it is imagined by the author of the book.

The personal contribution of the author consists in the development of models of distribution of the set reliability between elements of highly reliable system of ensuring anti-terrorist safety by the criterion of use of the minimum financial resource, models of distribution of a limited financial resource between elements of highly reliable system of ensuring anti-terrorist safety by the criterion of ensuring the maximum probability of protection, model of "the system of ensuring anti-terrorist safety of airline", model of assessment of the probability of realization of act of terrorism depending on quantity and reliability of boundaries of protection, development of a method of selection of the experts involved in ensuring anti-terrorist safety of airline in qualification qualities and also methodical recommendations about the selection of the experts involved in ensuring anti-terrorist safety of airline in personal qualities, to the preparation of the main scientific publications on the conducted research.

The practical value of the research conducted by the author is confirmed by the recommendations of the Interstate Aviation Committee (IAC), GOSOBORON-ZAKAZ (LLC), Russian Academy of National Economy and Public Service (RANEPA) under the president of the Russian Federation.

According to the author of the book, further steps in the field of decreasing the risk of realization of terrorist threats in airline should be directed to the creation of a complex information and analytical security system capable in real time to carry out assessment of the current state of anti-terrorist security of objects of aviation infrastructure and to define the most vulnerable elements, considering both indicators of reliability of technical security systems, and directly a role of a human factor—organizational and administrative measures and questions of staffing.

Dictionary of Terms

AB	Aviation safety;
ANTB	Anti-terrorist safety;
ANV	The act of illegal intervention;
AP	Aviation incident;
The EXPERT	An emergency;
ASKUD	The automated control and management access system;
BP	Safety of flights;
VS	The aircraft;
VV	Explosives;
W	Air transport;
VU	Explosive devices;
HECTARE	Civil aviation;
ICAO (ICAO)	International Civil Aviation Organization;
MAU	Moscow aviation knot;
Ministry of Internal Affairs	Ministry of Internal Affairs of the Russian Federation;
MFA	Ministry of Foreign Affairs of the Russian Federation;
Department of Internal Affairs	Internal Affairs Service Russian Federation (Police);
OTI	An object of transport infrastructure;
Rosgvardiya	Troops of national guard of the Russian Federation;
PS	The border service;
SAB	Service of aviation safety;
SPB	The systems of fire safety;
TB	Transport safety;
TSN	The television systems of observation;
TSO	Technical means of protection;

P. P. Hachikyan, *Countering Air Terrorism*, Springer Aerospace Technology,
https://doi.org/10.1007/978-981-19-6172-4

FSB (or FSS)	Federal Security Service of the Russian Federation;
The BSF	A human factor.
"*"	The indicated symbol in the text marks the names of organizations whose destructive activities are prohibited and illegal.

Appendix A

1. The scale "Crime" allows defining the tendency of the candidate to commission of crimes and asocial behavior (33 questions) (Table A.1).
2. The scale "The conscious expressed hostility" allows defining the tendency of the candidate to conscious hostility concerning people around (47 questions) (Table A.2).
3. The scale "Immorality" allows defining tendency of the candidate to immoral behavior (6 questions) (Table A.3).
4. The scale "Bias" allows defining the presence at the candidate of the created prejudices (32 questions) (Table A.4).
5. The scale of "The double standard of the personality" allows defining presence at the candidate of lines of duplicity (49 questions) (Table A.5).
6. The scale "Trustfulness" allows defining trustfulness of the candidate (68 questions) (Table A.6).

Table A.1 Scale "crime"

26 true	21, 26, 33, 38, 56, 116, 118, 143, 146, 223, 224, 254, 260, 298, 342, 355, 419, 421, 434, 458, 471, 477, 485, 537, 561, 565
7 false	37, 141, 173, 177, 294, 427, 464
Male	$M = 11.28; \sigma = 4.23$
Female	$M = 10.86; \sigma = 3.81$

Table A.2 Scale "the conscious expressed hostility"

42 true	16, 28, 35, 39, 75, SO, 93, 97, 109, 110, 117, 118, 121, 123, 127, 136, 139, 145, 197, 226, 233, 234, 235, 265, 269, 271, 280, 316, 336, 355, 381, 393, 417, 426, 437, 438, 447, 452, 469, 471, 504, 507
5 false	82, 96, 347, 399, 468
Male	$M = 16.4: \sigma = 6.12$
Female	$M = 17.42; \sigma = 5.59$

P. P. Hachikyan, *Countering Air Terrorism*, Springer Aerospace Technology,
https://doi.org/10.1007/978-981-19-6172-4

Table A.3 Scale "immorality"

5 true	143, 250, 271, 277, 298
1 false	289
Male	M = 2.54; σ = 1.39
Female	M = 2.72; σ = 1.36

Table A.4 Scale "bias"

29 true	47, 84, 93, 106, 117, 124, 136, 139, 157, 171, 186, 250, 280, 304, 307, 313, 319, 323, 338, 349, 373, 395, 406, 411, 435, 437, 469, 485, 543
3 false	78, 176, 221
Male	M = 10.46; σ = 4.39
Female	M = 11.4; σ = 4.27

Table A.5 Scale of "the double standard of the personality"

45 true	13, 26, 58, 94, 111, 112, 119, 129, 147, 158, 206, 232, 289, 317, 336, 337, 338, 356, 357, 361, 375, 378, 380, 390, 392, 395, 397, 402, 404, 413, 416, 439, 443, 457, 461, 468, 470, 492, 499, 502, 506, 509, 510, 548, 564
4 false	30, 45, 176, 401
Male	M = 19.2; σ = 6.32
Female	M = 4.37; σ = 6.05

Table A.6 Scale "trustfulness"

25 true	2, 36, 51, 95, 109, 153, 174, 181, 187, 192, 208, 221, 231, 234, 253, 270, 355, 367, 380, 410, 421, 430, 458, 513, 515
43 false	14, 22, 32, 33, 34, 43, 48, 58, 62, 82, 94, 100, 132, 140, 1S9, 209, 217, 236, 241, 244, 251, 261, 341, 344, 349, 359, 378, 384, 389, 420, 483, 488, 489, 494, 510, 525, 541, 544, 548, 554, 559, 561, 555
Male	M = 47.7; σ = 5.6
Female	M = 41.8; σ = 6.2

7. The scale "The relation to work" allows defining the candidate's relation to work (37 questions) (Table A.7).
8. The scale of "The devil of responsibility" allows defining level of responsibility of the candidate (63 questions) (Table A.8).
9. The scale "Social responsibility" allows determining the level of social responsibility of the candidate (32 questions) (Table A.9).
10. The scale "Altruism" allows determining altruism level at the candidate (9 questions) (Table A.10).

Table A.7 Scale "relation to work"

29 true	13, 16, 32, 35, 40, 41, 59, 84, 109, 112, 170, 244, 250, 259, 272, 301, 312, 331, 335, 343, 389, 395, 404, 406, 435, 487, 507, 526, 549
8 false	3, 9, 88, 164, 207, 257, 318, 407
Male	M = 11.67; σ = 3.92
Female	M = 14.66; σ = 4.22

Table A.8 Scale of "devil of responsibility"

63 true	6, 12, 69, 77, 79, 91, 93, 95, 99, 100, 102, 109, 111, 136, 141, 162, 165, 166, 181, 232, 240, 244, 248, 254, 255, 264, 270, 280, 292, 298, 304, 319, 321, 329, 340, 361, 367, 373, 391, 400, 410, 415, 416, 421, 425, 427, 429, 439, 444, 447, 455, 465, 475, 477, 489, 491, 492, 499, 500, 503, 510, 523, 564
0 false	–
Male	M = 33.58; σ = 4.87
Female	M = 35.05; σ = 6.28

Table A.9 Scale "social responsibility"

8 true	58, 111, 173, 221, 294, 412, 501, 552
24 false	6, 28, 30, 33, 56, 116, 118, 157, 175, 181, 223, 224, 260, 304, 419, 434, 437, 468, 469, 471, 472, 529, 553, 558
Male	M = 13.92; σ = 2.76
Female	M = 13.68; σ = 2.49

Table A.10 Scale "Altruism"

0 true	–
9 false	19, 26, 28, 80, 89, 112, 117, 120, 280
Male	M = 4.16; σ = 1.88
Female	M = 4.30; σ = 1.5

Appendix B

P. P. Hachikyan, *Countering Air Terrorism*, Springer Aerospace Technology,
https://doi.org/10.1007/978-981-19-6172-4

Questions (statements). Intentional method of personnel selection (candidate's experiences about the outside world = d)					
Scale					
0 — 0,1 — 0,2 — 0,3 — 0,4 — 0,5 — 0,6 — 0,7 — 0,8 — 0,9 — 1					
№	Statement	Answer options (values): 0; 0,1; 0,2; 0,3; 0,4; 0,5; 0,6; 0,7; 0,8; 0,9; 1			Statement
1	I am not sure that preventing a terrorist attack will improve the opinion of others about me	Candidate's numerical response (value) =			If I prevent a terrorist attack, everyone who knows about it will start treating me better
2	I believe that people around me cannot be interested in the concerns of other people in any way (they have enough of their own)	Candidate's numerical response (value) =			I believe that other people's (my) concerns may also be important to others
3	Others don't care how well I do my job	Candidate's numerical response (value) =			Others understand and appreciate the importance of my work
4	The surrounding people (including family, friends, relatives) do not notice me	Candidate's numerical response (value) =			There are at least a certain (at least 1) number of people who really need me
5	I wouldn't want to get better for those around me	Candidate's numerical response (value) =			I would like to be better for those around me
6	Society will not appreciate my actions if I save just one person	Candidate's numerical response (value) =			Society will appreciate my actions if I save even one person
7	Society values only heroes who have saved many human lives	Candidate's numerical response (value) =			Society values everyone who saved people's lives, no matter how many of them there were
8	Society and the state are unfair - heroes often remain without awards and glory, forgotten, even if they die	Candidate's numerical response (value) =			Real heroes (even if they are ready to die) do not need awards and glory - it is more important to realize the usefulness of their actions for society

9	I believe that the employer usually demands too much from the employee	Candidate's numerical response (value) =			I believe that the employer usually understands in good faith what to demand from the employee (how much/little)
10	I believe that it is possible to become famous in society and bad deeds (I like it)	Candidate's numerical response (value) =			I believe that you will not become famous for bad deeds in society (only good deeds impress me)
Questions (statements). Intentional method of personnel selection (candidate's experiences about the inner world = D)					
№	Statement	Answer options (values): 0; 0,1; 0,2; 0,3; 0,4; 0,5; 0,6; 0,7; 0,8; 0,9; 1			Statement
1	I am capable of breaking the law under certain circumstances	Candidate's numerical response (value) =			I am not capable of breaking the law under any circumstances
2	My friends can influence my opinion	Candidate's numerical response (value) =			My friends can't influence my opinion
3	There are no authorities who could influence me and encourage me to do certain things	Candidate's numerical response (value) =			There are people who know much more than me and can share their knowledge with me, but I have to clearly distinguish the bad from the good
4	If I am being blackmailed, I will try to negotiate and I am ready to do anything to settle everything without making too much noise	Candidate's numerical response (value) =			I do not succumb to blackmail under any circumstances, even if it discredits me and/or harms me
5	I am ready to do almost anything if I am well paid	Candidate's numerical response (value) =			I am ready to perform any actions for payment, but only if they are not directed against other people, the state and are morally permissible in society
6	I am ready for anything if I find out that the life of my loved ones depends on my decisive actions	Candidate's numerical response (value) =			I am not ready for everything, even if I know that the life of my loved ones depends on my decisive actions
7	Inside me there is a constant struggle between good and evil	Candidate's numerical response (value) =			Inside me, it has long been determined what is good and what is bad
8	Only I can solve all my problems by myself, no one will help me	Candidate's numerical response (value) =			I'm sure there are those who are willing to help me, even if they don't know me very well

9	I am convinced that if you don't want problems and you value your work (service) - the boss is always right	Candidate's numerical response (value) =			I believe that the boss is right when he does not violate the law and acts according to job descriptions
10	I am impressed by the expression of Karl Marx - "The liberation of the working class must be won by the working class itself"	Candidate's numerical response (value) =			I am not impressed by the expression of Karl Marx - "The liberation of the working class must be won by the working class itself"

Calculation of results on a scale (candidate's experiences about the outside world = d)		
№ statem.	Result (digital value)	
1	0	
2	0	
3	0	
4	0	
5	0	
6	0	
7	0	
8	0	
9	0	
10	0	
	Average result	0,000

Calculation of results on a scale (candidate's experiences about the inner world = D)		
№ statem.	Result (digital value)	
1	0	
2	0,0	
3	0	
4	0	
5	0	
6	0	
7	0	
8	0	
9	0	
10	0	
	Average result	0,000

Summing up (candidate's experiences about the outside world = d)	
The direction of the intention	Negative
Explanation: the point of separation of the intention into positive and negative, as an example, taking into account the theory of V.A. Lefevre, is given by the golden section of Fibonacci - the number 0.618. If the average result is >=0.618, the intention is positive. If the average result is <=0.617, the intention is negative.	

Summing up (candidate's experiences about the inner world = D)	
The direction of the intention	Negative
Explanation: the point of separation of the intention into positive and negative, as an example, taking into account the theory of V.A. Lefevre, is given by the golden section of Fibonacci - the number 0.618. If the average result is >=0.618, the intention is positive. If the average result is <=0.617, the intention is negative.	

The final result of the selection according to the intentional method of personnel selection	
The average total value of indicators on the scales (d,D)	0
Candidate	Not suitable
Explanation: taking into account the theory of V.A. Lefebvre, we will take the candidate's experiences about the outside world (d) as the previously specified variable x1 - a measure of pressure from the state and society, inclining the candidate to commit a legitimate act, while the candidate's experiences about the inner world (D) as his real intention to commit an act, the previously specified variable x3 - a measure internal aspiration (intention) of the candidate to commit a legitimate act. This makes it possible to bring both values of variables x1 and x3 to the decisive ones (1 is the positive pole, 0 is the negative pole) according to theory, i.e. to balance their significance for decision-making and provide an opportunity to obtain an overall average result on the scales, according to which to make a final decision on validity. At the same time, the numerical values of the answers to the questions on the d and D scales, respectively, can be used for a detailed study of the candidate's personality, beyond the automatic processing of the results.	

Pay attention (binary method)! For example and clarity, the author provides only a decoding of the use of the "Crime" scale, for the full use of the methodology, it is necessary to adapt the remaining another nine scales of the SMIL questionnaire to the study independently according to the given example.

Summarizing the results of the binary method of personnel selection for males							
№		Measure			Scale		
1.	x1	Pressure of the state and society, which inclines the candidate to commit a legitimate act	L1 =	1	"Social responsibility"		
2.	x2	Subjective assessment (by the candidate) of the pressure of the state and society towards the commission of a legitimate act	L2 =	0	h1 =	0	"Crime"
					h2 =	0	"Conscious expressed hostility"
					h3 =	0	"Immorality"
					h4 =	0	"Prejudice"
					h5 =	0	"Double standard of personality"
3.	x3	Internal aspiration (intention) of the candidate to commit a legitimate act	L3 =	0	k1 =	0	"The line of responsibility"
					k2 =	0	"Attitude to work"
					k3 =	0	"Altruism"
					k4 =	0	"Credulity"

Calculation of results on composite scales:
$L_2 = \frac{(h_1+h_2+h_3+h_4+h_5)}{z}$
, where z is the number of h elements (application – 5)
$L_3 = \frac{(k_1+k_2+k_3+k_4)}{e}$
, where e is the number of k elements (application – 4)

"The Formula of person" by V.A. Lefevre	
$X_1 = 1-(1-x_3+x_2x_3)+x_1(1-x_3+x_2x_3)$	
X1 =	1

Final selection result	
Candidate	Suitable

Summarizing the results of the binary method of personnel selection for females							
		Measure			Scale		
1.	x1	Pressure of the state and society, which inclines the candidate to commit a legitimate act	L1 =	1	"Social responsibility"		
2.	x2	Subjective assessment (by the candidate) of the pressure of the state and society towards the commission of a legitimate act	L2 =	0	h1 =	0	"Crime"
					h2 =	0	"Conscious expressed hostility"
					h3 =	0	"Immorality"
					h4 =	0	"Prejudice"
					h5 =	0	"Double standard of personality"
3.	x3	Internal aspiration (intention) of the candidate to commit a legitimate act	L3 =	1	k1 =	1	"The line of responsibility"
					k2 =	1	"Attitude to work"
					k3 =	0	"Altruism"
					k4 =	0	"Credulity"

Calculation of results on composite scales:

$$L_2 = \frac{(h_1 + h_2 + h_3 + h_4 + h_5)}{z}$$

, where z is the number of h elements (application – 5)

$$L_3 = \frac{(k_1 + k_2 + k_3 + k_4)}{e}$$

, where e is the number of k elements (application – 4)

"The Formula of person" by V.A. Lefevre

$$X_1 = 1 - (1 - x_3 + x_2 x_3) + x_1 (1 - x_3 + x_2 x_3)$$

X1 =	1

Final selection result	
Candidate	**Suitable**

Additional scale of the SMIL questionnaire "Crime" (data for the survey of male)		
Negative answers are "True":		
№ by SMIL	The content of the question on the SMIL questionnaire	Answer (true - 1, false - 0)
21	At times I really wanted to leave the house.	0
26	If I get into an unpleasant situation, then it's best for me to keep my mouth shut.	0
33	I have strange and peculiar experiences.	0
38	There was a time in my youth when I committed petty thefts.	0
56	When I was a boy, I was kicked out of school for absenteeism.	0
116	I get more pleasure from playing or racing if I bet.	0
118	At school, I was sometimes called to the principal for absenteeism.	0
143	As a child, I was in a company where everyone stood up for each other in everything.	0
146	I have a passion for changing places, and I can't be happy without traveling and wandering.	0
223	I really love hunting.	0
224	My parents often disapproved of my dating.	0
254	I like to be in a company where they like to make fun of each other.	0
260	At school, I did not assimilate the material well.	0
298	When several people get into an unpleasant story, it's better for them to agree on what to say, and then all stick to the same line.	0
342	I immediately forget what they tell me.	0
355	Sometimes it feels good to hurt those I love.	0
419	I often skipped school.	0
421	There are some very nervous people in my family.	0
434	I would like to be a racing driver.	0
458	The man who was most involved with me as a child (father, stepfather or another person) was strict with me.	0
471	My grades at school were systematically bad.	0
477	If I got into trouble with my friends, guilty to the same extent as I am, I would rather take all the blame on myself than give them away.	0
485	A man, being in the company of a woman, usually thinks about things related to her gen	0
537	I would like to hunt lions or tigers.	0
561	I really like horse riding.	0
565	I feel like jumping down when I'm at a height.	0
	Total points:	0

Additional scale of the SMIL questionnaire "Crime" (data for the survey of female)		
Negative answers are "True":		
№ by SMIL	The content of the question on the SMIL questionnaire	Answer (true - 1, false - 0)
21	At times I really wanted to leave the house.	0
26	If I get into an unpleasant situation, then it's best for me to keep my mouth	0
33	I have strange and peculiar experiences.	0
38	There was a time in my youth when I committed petty thefts.	0
56	When I was a boy, I was kicked out of school for absenteeism.	0
116	I get more pleasure from playing or racing if I bet.	0
118	At school, I was sometimes called to the principal for absenteeism.	0
143	As a child, I was in a company where everyone stood up for each other in eve	0
146	I have a passion for changing places, and I can't be happy without traveling and wandering.	0
223	I really love hunting.	0
224	My parents often disapproved of my dating.	0
254	I like to be in a company where they like to make fun of each other.	0
260	At school, I did not assimilate the material well.	0
298	When several people get into an unpleasant story, it's better for them to agree on what to say, and then all stick to the same line.	0
342	I immediately forget what they tell me.	0
355	Sometimes it feels good to hurt those I love.	0
419	I often skipped school.	0
421	There are some very nervous people in my family.	0
434	I would like to be a racing driver.	0
458	The man who was most involved with me as a child (father, stepfather or another person) was strict with me.	0
471	My grades at school were systematically bad.	0
477	If I got into trouble with my friends, guilty to the same extent as I am, I would rather take all the blame on myself than give them away.	0
485	A man, being in the company of a woman, usually thinks about things relate	0
537	I would like to hunt lions or tigers.	0
561	I really like horse riding.	0
565	I feel like jumping down when I'm at a height.	0
	Total points:	0

Negative responses are "False":			
№ by SMIL	The content of the question on the SMIL questionnaire		
37	I have never been in trouble because of my sexual behavior.		0
141	My behavior is largely determined by the rules and habits of the people around me.		0
173	I loved school.		0
177	My mother is a good woman.		0
294	I've never broken the law.		0
427	I am confused by indecent jokes.		0
464	I am deprived of the ability to evoke any visions in myself.		0
		Total points:	0
		Total points total:	**0**

The result obtained is "raw" T-scores (X)	0

Statistical scientific data on SMIL for the formula		
For male:		
M is the average normative indicator (median)		11,28
σ - the standard deviation		4,23

Formula for calculating T-points by SMIL (L.Sobchik) for men (m = male)

$$\gamma = 50 + \frac{10(X_m - M_{m.})}{\sigma_m}$$

γ = 23

Negative responses are "False":			
№ by SMIL	The content of the question on the SMIL questionnaire		
37	I have never been in trouble because of my sexual behavior.		0
141	My behavior is largely determined by the rules and habits of the people around me.		0
173	I loved school.		0
177	My mother is a good woman.		0
294	I've never broken the law.		0
427	I am confused by indecent jokes.		0
464	I am deprived of the ability to evoke any visions in myself.		0
		Total points:	0
		Total points total:	**0**

The result obtained is "raw" T-scores (X)	0

Statistical scientific data on SMIL for the formula		
For female:		
M is the average normative indicator (median)		10,86
σ - the standard deviation		3,81

Formula for calculating T-points by SMIL (L.Sobchik) for men (f = female)

$$\gamma = 50 + \frac{10(X_f - M_f)}{\sigma_f}$$

γ = 21

Technically acceptable interval (Min. - Max.), in points	
T-points Min.	23
T-points Max.	101
The most suitable result (to the best), the Fibonacci golden ratio is 1.618	37
Acceptable deviation is the selection interval (set by the top step of the hierarchy), in points	10
The "numerical expiration date" of the candidate, in points	
From	27
To (before)	47
Interpretation according to the "Person formula" by V.A. Lefevre	
According to the principle - a positive candidate if he falls into the interval (fit) - 1; a negative candidate if he does not fall into the interval (not fit) - 0.	0
The result of the selection on the scale	**Not suitable**

Technically acceptable interval (Min. - Max.), in points	
T-points Min.	21
T-points Max.	108
The most suitable result (to the best), the Fibonacci golden ratio is 1.618	34
Acceptable deviation is the selection interval (set by the top step of the hierarchy), in points	10
The "numerical expiration date" of the candidate, in points	
From	24
To (before)	44
Interpretation according to the "Person formula" by V.A. Lefevre	
According to the principle - a positive candidate if he falls into the interval (fit) - 1; a negative candidate if he does not fall into the interval (not fit) - 0.	0
The result of the selection on the scale	**Not suitable**

Printed by Printforce, the Netherlands